Practical Animal Nutrition

Practical Animal Nutrition

(as per new VCI syllabus ANN-111, ANN-121, ANN-211)

Dr. Vishal Mudgal
Assistant Professor / Scientist
Department of Animal Nutrition
College of Veterinary Science and Animal Husbandry
Jabalpur-482 004 (M.P.)

2012

nipa

New India Publishing Agency
Pitam Pura, New Delhi-110 088

Published by
Sumit Pal Jain *for*
New India Publishing Agency
101, Vikas Surya Plaza, CU Block, L.S.C. Mkt.,
Pitam Pura, New Delhi-110 088, (India)
Ph.: 011-27341616, Fax: 011-27341717, Mob.: 09717133558
E-mail: info@nipabooks.com
Web: www.nipabooks.com

ISBN : 978-93-80235-97-4

Typeset at: Harminder *for* Laxmi Art Creation
Printed at: Jai Bharat Printing Press, Delhi

Preface

There are several textbooks on Theoretical aspect of Animal Nutrition written by eminent scientists. But no single textbook is their till the date regarding the practical courses of the subject for undergraduate students of Veterinary science by providing all the needed information at single place. As per the Veterinary Council of India regulations now Animal Nutrition course has been redesigned into three courses for the first three semesters of the degree and including the courses on Principles of Animal Nutrition and Feed Technology (ANN-111), Applied Animal Nutrition-I (ANN-121), dealing with ruminants animals and last on is Applied Animal Nutrition-II (ANN-211), deals with Non-ruminants animals. As per the new VCI syllabus the book covers all the three courses in three different parts with different chapters covering all the topics of the practical.

Jabalpur (M.P.)
20 June, 2011

Author

Contents

Part-II : Applied Animal Nutrition (Ruminants, ANN-121)

Part-III : Applied Animal Nutrition (Non-Ruminants, ANN-211)

Part I

Principles of Animal Nutrition and Feed Technology (ANN-111)

Exercise No. 1

Familiarization of Various Feedstuffs, Fodders and their Selection

Livestock feeds are generally classified according to the amount of a specific nutrient they furnish in the ration. They are divided into two general classes roughages and concentrates. Roughages are bulky feeds containing relatively large amount of less digestible material, i.e. crude fiber more than 18% and lower than 60% TDN on dry matter basis. Concentrates are feeds which contain relatively smaller amount (less than 18%) of fiber and have comparatively high digestibility and as a result higher nutritive value having more than 60% TDN.

A. Roughages

Roughages are subdivided into two major groups - succulent and dry, based upon the moisture content. Succulent feeds usually contain moisture content from 60 to 90%, whereas dry contains only 10 to 15% moisture. Succulent may again be divided into pasture, cultivated fodder crops, tree leaves, silage and root crops. Dry roughages may be further classified as hay and straw based on the nutritive value and methods of preservation.

Succulent roughages

1. *Pasture*: Of the succulent feeds, pasture is the most convenient and economic for maintaining large livestock. Young rapidly growing grasses are rich in protein and highly palatable.

2. *Cultivated fodder crops*: These may be classified into leguminous and non-leguminous. Among non leguminous fodder- jowar, maize and sudan grass are the most common kharif fodder and rabi fodder includes oats and barley. A non leguminous perennial fodder crop consists of Napier grass, Hybrid napier grass, guinea grass and para grass. Among the leguminous fodder group cowpea and guar are the kharif crops and rabi crops include the Berseem, which is annual crop and Lucerne is the perrineal crop.
3. *Tree leaves*: These are usually used in the ration of sheep and goats. It includes the leaves of jharberi, katchnar, peepal, babul and bel, which may serve as maintenance type ration.
4. *Root crops*: These crops are rich in moisture content and low in fiber content. It includes turnips, carrot and tapioca *etc.*
5. *Silage*: This is a conserved crop usually used in the period of scarcity, when greens are lacking to feed animals. The crops mainly used for silage preparation is rich in carbohydrates like maize, jowar, bajra *etc.*

Dry roughages

1. *Hay*: This is also a conserved forage and it is prepared just by reducing the moisture content of green crop up to suitable level (15% or lesser) at which the crop may be stored for a longer period to be used in adverse climate when greens are not available. Hay is usually prepared from crops which may be easily dried without affecting its nutritive value adversely. The crops commonly used are of Lucerne and grasses.
2. *Straws*: These are the by-product of thrashing seeds and are usually very poor nutritionally. They are rich in crude fiber and leguminous straws are superior nutritionally over non leguminous straws. Leguminous straw includes of cowpea, pea, ground nut, mung, gram etc, while in non leguminous wheat straw and paddy straw are very common.

B. Concentrates

These are the feeds which supplies primary nutrients (proteins, carbohydrates and fats) at higher level thus TDN more than 60%, but contain less than 18% crude fiber.

The concentrates may be further divided on the basis of crude protein (CP) content into energy rich concentrates, when CP content is less than 18% or protein rich concentrates, when the CP value exceeds 18%.

1. Energy rich concentrates: It includes grains and seeds, mill by-products, molasses and roots.
   A. *Grains and seeds*: It includes maize, barley, oats, sorghum, bajra.
   B. *Mill by-products*: It includes brans (Rice bran, wheat bran and maize bran), flour, germ, gluten, grain screenings, grouts, hulls, malt sprouts, meal, middlings, polishing, red dog and shorts.
   C. *Molasses*: these are rich in sugars, hemicelluloses and minerals. It includes cane molasses, beat molasses, citrus molasses and wood molasses.
   D. *Root and tubers*: Root crops are high in moisture (70-90%) and low in fiber (5-12%) content. They are rich in sugars and low in crude protein. It includes potatoes, cassava and turnips etc.
2. Protein rich concentrates: They may be subdivided according to source of origin like- Plant proteins, Animal proteins, Non protein nitrogen and single cell protein.
   A. *Plant proteins*: It includes oil seed meals like ground nut or peanut oil meal, linseed meal, mustard or sarson cake, cotton seed cake, coconut meal, Til cake or sesame meal, soybean oil meal, brewer's grains etc.
   B. *Animal protein*: It includes meat meal or meat scrape, meat and bone meal or scrap, blood meal, feather meal, hatchery by-product meal, fish meal etc.
   C. *Non protein nitrogen feedstuffs*: Feedstuffs which contains nitrogen in a form other than proteins or poly-peptides are termed as non protein nitrogen. It includes urea, biurate and ammonium salts etc.
   D. *Single cell protein*: It obtained from single cell organisms like yeast, bacteria and algae.

C. Mineral supplements

It includes natural or pure elements like oyster shell meal, ground lime stone, bone meal, dicalcium phosphate, deflorinated rock phosphate, common salt *etc.*

D. Vitamin supplements

It includes natural or pure vitamins.

E. Additives

It includes antibiotics, hormones, coloring agents *etc.*

Exercise No. 2

Preparation and Processing of Various Samples for Analysis

The procedure used in the preparation of samples depends on the purpose of analysis and the nature of the constituents that are to be determined.

a. Plant material (Herbage)

1. Separate the plant material by using sharp knife.
2. Remove the contaminations like soil or other material by gentle brushing. Never wash the plant material with water.
3. Mix corresponding plant parts from more than one plant to obtain representative sample of sufficient size for analysis.
4. Cut the sample immediately into small pieces and mix thoroughly.
5. Weigh a portion of this sample for those analyses which are to be performed on green material.
6. Dry the sample in a hot air oven at 100 ± 2°C for storage.

b. Cattle feaces

1. Mix daily collection thoroughly on a clean surface by hand.
2. Sub-sample - 1% of wet weight for dry matter estimation.

3. Sub-sample - 0.1% of wet weight for nitrogen determination and preserve in bottle with 40% (w/v) H_2SO_4 after through mixing.

c. Sheep / goat feaces

Mix the feaces pellets thoroughly and take sub-samples of 5 and 1% for dry matter and nitrogen analysis respectively, as discussed earlier.

d. Poultry excreta

1. Collect excreta on polythene sheet.
2. Separate feathers, feed and other contaminations.
3. Mix thoroughly and take sub samples of 15 and 5% separately for dry matter and nitrogen determination, respectively. For nitrogen analysis, samples should be preserved with 5% (w/v) sulphuric acid.
4. Preserve dried and ground samples for further analysis.

e. Urine

The samples of urine (5-20% depending upon the size of the animal) are collected daily in bottles and evaporation of ammonia N is prevented by adding 5% (w/v) H_2SO_4.

f. Silage

1. Sample of silage is collected from different parts of silo pit. Cut into pieces of 3 to 5 cm length with the help of a hand chopper on clean surface. Then chopped silage is mixed thoroughly and about 500g material is taken in a numbered polythene bag and closed immediately.
2. Weigh fresh sample for the analysis of nitrogen and dry matter.
3. Prepare silage extracts for determination of the pH, volatile fatty acids, ammonia nitrogen *etc.*
4. Preserve dried ground samples for determination of different proximate principles.

g. Hay and straw

Take about 1 kg of straw from various places of the heap. Chaff the sample, dry it and grind in a mill through 1mm sieve and preserve in plastic bag for analysis.

h. Concentrate feed

Take representative samples from different parts of each bag. Mix the sample thoroughly and determine its dry matter content. Grind the sample and preserve for further analysis.

The following are the approximate quantities of samples of feed taken for different fractions of proximate principals:

Parameters	**Dry roughages (g)**	**Grains /grains by products (g)**	**Oil seed cakes (g)**	**Animal by products (g)**	**Green fodders (g)**
Moisture	10-15	10-15	5-10	5-10	50-100
Crude protein	2-3	1.5-2	0.5-1.0	1-2	10-12
Crude fat	3-5	2-3	1.5-2	2-3	2-3
Crude fibre	3-5	2-3	1.5-2	2-3	2-3
Total ash	5-10	5-10	5-10	5-10	5-6

Exercise No. 3

Proximate Principles or Weende's System of Analysis

Two German scientists Henneberg and Stohmann working at the Weende Experimental Station, Germany devised proximate principles analysis system in 1860's for routine description of animal feedstuffs. Therefore this system is also known as Weende system of analysis. Proximate analysis is a system of expressing the nutrient composition of a feed. The principle of analysis is to separate the nutrient present in the feedstuffs into various fractions, as proximate principles, which are six i.e. moisture, ash, crude protein, ether extract, crude fibre and nitrogen free extract.

Water (Moisture)

The moisture is determined as the loss in weight which results from drying a known weight of feed to constant weight at 100°C. Therefore, dry matter is estimated as a part of substance that does not evaporate at 100°C. The major difference in nutritional value on as fed basis is traceable to moisture content and dry matter. Calculation of relative cost of nutritional value involves consideration of moisture content.

Determination of moisture is essential in bulk purchase of feed ingredients, when the grain crop is new; the moisture content is normally high. In storing, the determination of moisture in feeds is very important. Feeds containing more than 14 percent moisture should not be stored in bulk for the danger, as spontaneous combustion.

Crude protein

The crude protein content is calculated from the nitrogen content of the feed and knowing the protein content of a feed provides an idea about the class of feed it belongs to. In this method, the feed is digested with sulphuric acid which converts all nitrogen to ammonia. This ammonia is liberated by the addition of sodium hydroxide to the digest, distilled off and collected in standard acid. The quantity so collected is determined by titration. Determination of crude protein is involves multiplication of estimated nitrogen content by a factor of 6.25, based on the assumption that all the protein contains 16% nitrogen and all nitrogen is present in the form of proteins. However, both the assumptions are not true. The term crude protein does not distinguish the nitrogen contribution from true protein and non-protein nitrogenous substances like urea, uric acid, free amino acids, ammonium salts etc. Non-protein nitrogen can be calculated by subtracting true protein content from crude protein.

Ether extract

The ether extract or crude fat is determined by subjecting the feed to a continuous extraction with petroleum ether (boiling point = 40°-60°C or 60°-80°C) for a definite period. The residue after the evaporation of solvent is called ether extract. Ether extract is a source of energy and provides essential fatty acids. In addition to fat, ether extract contains waxes, organic acids, alcohols, pigments etc. and thus also known as crude fat.

Crude fiber

The carbohydrate of the feeds is contained in two portions, the crude fiber and nitrogen free extract. Crude fiber is determined by subjecting the residual feed free from moisture and ether extract to successive treatments with boiling acid and alkali of fixed concentration, the organic residue is crude fiber. Crude fiber contributes to the bulkiness of feed. The crude fiber fraction contains cellulose, hemicellulose and lignin in variable proportions. Originally the crude fiber was intended to represent the indigestible portion of feed; however, a large part of it may in fact be digested by ruminants. Crude fiber estimation simulates digestion in monogastric animals than in ruminants.

Nitrogen free extract

Nitrogen free extract is obtained when a sum of moisture, ether extract, crude fibre, crude protein and ash (per cent) is subtracted from 100. The value of nitrogen free extract is affected by analytical errors of these five parameters as well as lack of precision in crude fibre determination in separating functional categories of carbohydrates. The NFE content is inversely related to protein in concentrate feeds.

Total ash

The ash content is determined by ignition of a known weight of feed at 550-600°C. The residue obtained is ash and represents inorganic constituents of feeds. The ash, however, does not discriminate between the proportion of mineral matter and sand/silica due to either contamination or adulteration. The ash may contain materials of organic origin such as sulphur and phosphorus of proteins. Likewise some loss of sodium, chloride, selenium and iodine takes place on ignition. Indirectly, ash helps in calculating total carbohydrates, NFE and organic matter by difference.

Significance of weende analysis

1. It is a fundamental basis for the nutritional evaluation of feeds in laboratory.
2. It is a common basis for feed purchase and for ration formulation.
3. It is the starting point for more detailed analysis.

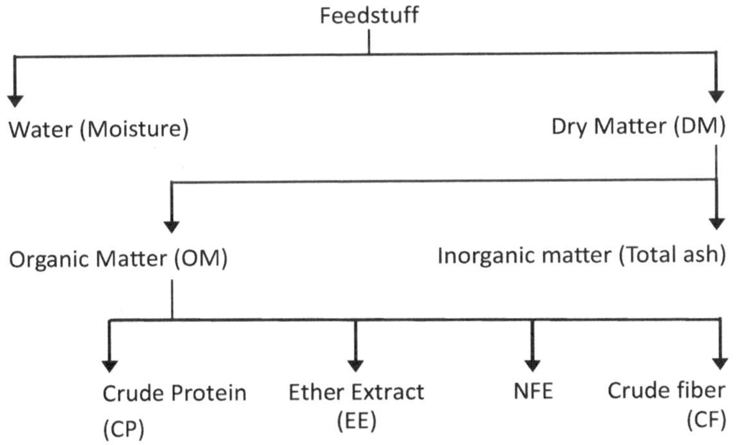

Weende's System of Chemical Analysis

Exercise No. 4

Estimation of Moisture and Dry Matter (DM)

Principle

If a sample of feed is heated and dried in a hot air oven at the temperature of boiling water (100°C ± 2°C), the water present in feed sample escapes as vapour. The loss in weight is considered as moisture and the residue is termed as dry matter (DM).

Requirement

1. Aluminium/steel moisture cup or Iron trey
2. Analytical balance
3. Dessicator
4. Hot air oven
5. Metal tongs
6. Spatula

Procedure

1. Dry the aluminium or stainless steel moisture cup (dish) in oven at 100°C for overnight, place in dessicator, cool and weigh.
2. Weigh in duplicate 2-10 g of sample into aluminium cup and shake the cup until the contents are evenly distributed.

3. Place the cup with lid removed in the oven and dry the contents for 12-18 hours (overnight) at 100° ± 2 °C.
4. Place cover on cup and transfer to dessicator to cool. Weigh as quickly as possible. Repeat the process of heating, cooling and weighing until the difference between two successive weighing is less than one milligram.
5. Report loss in weight as moisture and it can be calculated using the following formula.

Hot Air Oven

Calculation

Moisture % = 100 X Loss of moisture / Weight of sample
= 100 X (B - C) / (B – A)

Where,

A = Weight of empty moisture cup
B = Weight of moisture cup + sample
C = Weight of moisture cup + sample after drying
DM % = 100 – Moisture %

Exercise No. 5

Estimation of Total Ash and Organic Matter (OM)

Principle

Total ash is the inorganic residue which remains after a feed stuff is ignited to carbon free at 550-600°C in a muffle furnace. In other words, ash is that part of the material which does not disappear at 550-600°C, while all the organic matter (OM) gets burnet.

Requirement

1. Analytical balance
2. Asbestos sheet
3. Dessicator
4. Heater /gas flame burner
5. Metal tongs
6. Muffle furnace
7. Silica basin/crucibles

Procedure

1. Weigh in duplicate 2-10 g of feed sample depending upon the type of feed into a pre-weighed silica basin/crucible.

2. To carbonise the contents, burn and char the organic matter by placing the basin with sample on a low gas flame or heater until all smoke and black portions escape out. Do not stir or disturb the material while charring as this will result in a loss of the substance.

Muffle furnace

3. Keep the crucible in a muffle furnace and raise the temperature up to 550-600°C slowly by setting the regulator and complete the ignition of sample for 2-3 hours at the constant temperature.

4. Remove the silica basin/crucible holding with metal tong from muffle furnace and cool in dessicator and weigh.

5. Since ash is highly hygroscopic, note the lowest weight and report the total ash as per the calculation shown below:

Calculation

Total ash (%) on as such basis

= 100 X Weight of Total ash/ Weight of sample

= 100 X (C - A) / (B - A)

Where,

A = Weight of crucible

B = Weight of crucible + sample

C = Weight of crucible + Total Ash (weight of crucible after ashing)

Total ash (%) on DM basis = 100 X Wt of Total ash / Wt of sample taken on DM basis

Organic Matter % (OM) = 100 - Total ash %

Silica basin with ash can be preserved for estimation of acid insoluble ash and individual elements from the mineral extract.

Exercise No. 6

Estimation of Acid Insoluble Ash (AIA)

Principle

The portion of the total ash which is insoluble in dilute hydrochloric acid is called acid insoluble ash (AIA) and it is the part of ash which is having unavailable mineral matter like silica and sand. Silica is the part of mature plant materials which are highly lignified or woody like straw, husk, stems and stalks, while sand inters in feeds by faulty handling in the fields or as an adulterant.

Equipment/glass wares

1. Asbestos sheet
2. Filter paper (Whatman No.1)
3. Funnel stand
4. Glass funnel
5. Glass rod
6. Hot plate/heater/Bunsen flame
7. Metal tongs
8. Silica basin (crucible)
9. Volumetric flask
10. Wash bottle with distilled water

Reagents

1. Hydrochloric acid (1:1)

Procedure

1. After estimating the total ash, moisten the ash of basin with 5-10 drops of distilled water and then add about 20 ml of dilute HCl (1:1) along the sides of basin with the use of glass rod, now mix well and try to dissolve ash in acid and transfer all the ash quantitatively in to a 250ml beaker. Now heat the beaker over a Bunsen flame or hot plate/heater and boil for 5 to 10 minutes.
2. Cool and filter through a Whatman filter paper (No.1) into volumetric flask (250 ml). Wash silica basin several times with small quantities of hot distilled water. Transfer all the insoluble residue left in the silica crucible to the filter paper. Collect all the filtrate obtained in the washing into the volumetric flask and make the volume by adding distilled water up to the mark. This solution is called as acid extract or mineral extract and having all the essential minerals, if required to be estimated from it.
3. Transfer the filter paper containing the acid insoluble ash to the same silica basin used in the experiment after folding it into a small mass. Place the basin in a hot air oven at 100 ± 2°C for 1- 2 hour. Ignite it in a muffle furnace at 550-600°C for 2 hour after making it smokeless.
4. Remove the basin from muffle furnace and cool it in a dessicator and weigh. Record the lowest weight and calculate the acid insoluble ash as per the formula given on next page:

Calculation

Acid insoluble ash (% on as such basis)

= 100 x Weight of AIA/Weight of sample

= 100 x (C - A) / (B - A)

Where,

A = Weight of empty silica crucible

B = Weight of silica crucible + sample

C = Weight of silica crucible + acid insoluble ash

Acid insoluble ash (%, DM basis) = 100 x Wt. of AIA/Wt of sample on DM basis.

Exercise No. 7

Estimation of Crude Protein (CP)

Principle

Crude protein constitutes a mixture of true protein and non protein nitrogenous substances and is determined by estimating nitrogen and then multiplying the nitrogen content of feed by the factor 6.25. The factor 6.25 is derived on the basis that all proteins contain 16% nitrogen.

The total nitrogen content in the feedstuff is determined by Kjeldahl method which generally consists of three steps, i.e. digestion, distillation and titration. The sample is digested with an excess quantity of concentrated sulphuric acid which hydrolyses the protein, decomposes to ammonia and finally the ammonia combines with the excess of sulphuric acid forming ammonium sulphate. The acid digest is cooled, diluted with distilled water and distilled with strong alkali (sodium hydroxide, 40 per cent), the ammonia released during distillation is taken up into a boric acid solution forming ammonium borate complex and such ammonia is titrated with standard sulphuric acid. While estimating the nitrogen content, following reactions take place:

Digestion

$$\text{Protein} + H_2SO_4 \longrightarrow CO_2 + SO_2 + NH_3$$

$$2\,NH_3 + H_2SO_4 \longrightarrow (NH_4)_2SO_4$$

Distillation

Ammonium sulphate derived from feed nitrogen is converted into ammonia by adding 40 per cent alkali. Ammonia is received in boric acid solution by steam distillation.

$$(NH_4)_2SO_4 + 2NaOH \longrightarrow 2NH_3 + Na_2SO_4 + 2H_2O$$

$$NH_3 + H_2O \longrightarrow NH_4OH$$

$$NH_4OH + H_3BO_3 \longrightarrow NH_4B(OH)_4$$

Titration

The ammonia boric acid complex (ammonium borate) is titrated with standard H_2SO_4.

$$2NH_4B(OH)_4 + H_2SO_4 \longrightarrow (NH_4)_2SO_4 + 2H_3BO_3 + 2H_2O$$

Equipment/glass wares

1. Analytical balance
2. Beaker (100 ml or 250 ml)
3. Burette
4. Conical flask (100 ml)
5. Digestion bench
6. Glass beads
7. Glass rods
8. Kjeldahl flask (500 ml)
9. Macro/Micro Kjeldahl Distillation Assembly
10. Measuring cylinder
11. Metal tongs
12. Spatula
13. Wash bottle

Reagents

1. *Digestion mixture*: (9 parts potassium sulphate/sodium sulphate + 1 parts copper sulphate). Potassium sulphate/sodium sulphate raises the boiling point of the contents for efficient oxidation of material. Copper sulphate is added as a catalyst to fasten the reaction.

2. Boric acid/mixed/Toshiro's indicator: 10 ml of 0.1 per cent methyl red indicator + 5 ml of 0.1-0.5% bromocresol green indicator + 1000 ml of 2-4 per cent boric acid solution. For 0.1 per cent methyl red, dissolve 100 mg of methyl red in 100 ml of ethanol (95%). For 0.1-0.5 per cent bromocresol green, dissolve 100 mg -500 mg of bromocresol green in 100 ml of ethanol (95%). For 2-4 per cent boric acid, dissolve 20-40 g of boric acid in one litre of hot distilled water, cool and allow maturing for 2-3 days before decanting the clear liquid.
3. NaOH solution (40 per cent): Dissolve 400g of NaOH pellets in one litre of distilled water, cool and then use.
4. Conc. H_2SO_4 (AR)
5. Standard H_2SO_4 solution (N/7 or N/10 or N/70 or N/100)

Procedure

This method of nitrogen and crude protein determination is applicable for feed, faeces and meat samples. Macro/Micro Kjeldahl nitrogen analysis comprises three different stages viz., digestion, distillation and titration.

Digestion

1. Weigh 1-2 g of sample on a piece of butter paper depending upon the protein content and transfer it into a clean and dried Kjeldahl flask.
2. Add 5-10 g of digestion mixture to the sample taken in flask.
3. Add 20-30 ml of conc. H_2SO_4 along the neck of the flask so as to wash down any particles of the sample/digestion mixture sticking to the sides (increase the amount of H_2SO_4 by 10 ml for each gram of sample).
4. Drop a few glass beads into the flask to prevent spurting while heating.
5. Place the flask in an inclined position on a digestion heater and heat it below the boiling point of the acid until frothing ceases. If the froth of a sample starts up the neck of flask, remove it from the heater to allow froth to subside and return to heater. Increase heat until the acid boils vigorously and digests for a time after mixture is clear or until oxidation is complete (about 2 to 4 hours).

If any material sticks to the neck of the flask than cool the flask and wash down the black partials stick on the neck with a small amount of distilled water and redigest.

6. When the end point of digestion is reached, remove the Kjeldahl flask and allow it to cool.

A blank digestion should always be conducted simultaneously using all reagents in the same quantities except sample to be tested.

Distillation

Distillation process accomplished in Kjeldahl's distillation apparatus

1. Add about 25 ml of distilled water to the clear solution of flask, shake the flask for proper mixing and cool the flask under running tap water because the flask becomes very hot during mixing of contents with added water. Now transfer the digested feed sample quantitatively into a appropriate volumetric flask (100/250ml) and make the volume with water.
2. Take 10 ml of boric acid indicator into a conical flask/beaker and place it below the delivery tube of the distillation apparatus.
3. Add a representative digested sample (about 1-2ml) from volumetric flask in the chamber of distillation assembly and then add carefully NaOH solution (40 per cent) by the side of flask in quantity (say 5-10 ml) which is sufficient to make the solution alkaline so that it does not mix at once with the acid solution but forms a layer below the acid layer and then add 5 ml of distilled water to the same receiver for rinsing.
4. Take care that the tip of dip tube extends below the surface of boric acid solution otherwise ammonia would escape and affect the nitrogen content of sample. Ensure that the water runs in the condenser.
5. Heat the alkaline liquid by passing steam into it through the boiling distilled water in a flask. Which will remains continue until all ammonia passes over into the boric acid indicator. Faint reddish colour of boric acid starts changing to green colour since the released ammonia is absorbed by boric acid indicator. Released ammonia is allowed to distil till the original volume of boric acid indicator (10 ml) becomes three times more (30 ml).

When the distillation is practically complete, remove the conical flask/beaker containing distilled solution after washing the condenser tip thoroughly. Titrate the solution.

6. Distil the digested solution of blank flask in the same way so as to correct the nitrogen content on account of reagents used.

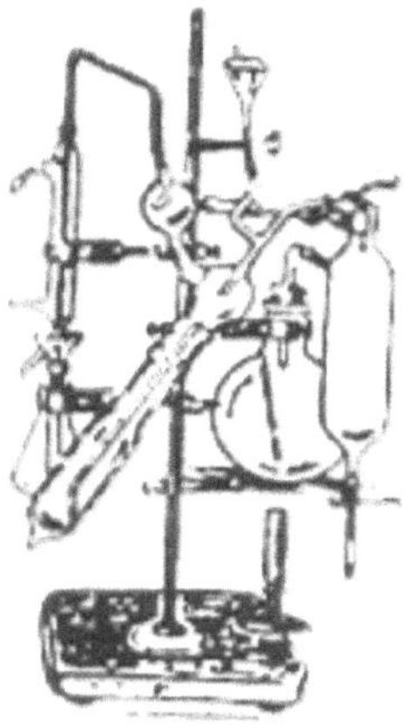

Distillation apparatus

Titration

Titrate the distillate (ammonia dissolved in boric acid indicator) with N/10 or N/7 standard sulphuric acid solution in macro Kjeldahl method and N/70 or N /100 standard sulphuric acid solution in micro Kjeldahl method. The end point will be again, red colour.

Calculation

Two atoms of nitrogen (in the form of ammonia) neutralise one molecule of sulphuric acid and form ammonium sulphate. Therefore, each ml of 1N sulphuric acid neutralised by ammonia is equivalent to 0.014 g or 14 mg nitrogen. Amount of nitrogen and crude protein per ml of sulphuric acid of different strengths may be noted from the following details.

H_2SO_4		Nitrogen		Crude protein	
Volume (ml)	Normality (N)	g	mg	g	mg
1	1N	0.0140	14	0.0875	87.5
1	N/10	0.0014	1.4	0.0875	8.75
1	N/100	0.00014	0.14	0.00875	0.875
1	N/7	0.002	2	0.0125	12.50
1	N/70	0.0002	0.2	0.00125	1.25

Nitrogen (%)

= 0.014 X ml of H_2SO_4 used X N X V X 100 / Wt of sample X A

Where,

N = Normality of the standard H_2SO_4 used

V = Volume made up of digested sample

A = Aliquot taken for distillation

For conversion of nitrogen into crude protein

CP (%) = N (%) x 6.25

CP = crude protein (%) content of feedstuff

N = total nitrogen (%) present in feedstuff

Exercise No. 8

Estimation of Ether Extract (EE)

Principle

When a sample of feed, which is free from moisture, is extracted continuously with fat solvent like petroleum ether, the vapour from the ether boiling in the flask passes into the condenser where it condenses and drops back on the sample which dissolves ether soluble materials such as fat, sterols, lecithin, resins and volatile oils in the extraction flask. The extract is termed as crude fat or ether extract (EE) because it also contains other fat like substances in addition to true fat.

Equipment/glass wares

1. Analytical balance
2. Beaker
3. Cotton wool
4. Dessicator
5. Extraction thimble
6. Hot air oven
7. Oil flask (150-250 ml)

8. *Soxhlet extraction assembly*: This consists of 3 parts (a) condenser at the top (b) Soxhlet or extractor in the middle (c) receiver flask or oil flask at the bottom. These three parts are assembled by means of their ground glass joints.
9. Heating mantle, provision for water supply to the condenser and electricity to heating mantle is also to be made.

Reagent

Petroleum ether (boiling point = 40-60°C or 60-80°C).

Procedure

1. Weigh 2-5 g of ground moisture free feed material on a butter paper, transfer it into an ready made extraction thimble or prepare it by whatman filter paper no 1 with support of test tube and get tightened with thread and plug the mouth of thimble with a piece of cotton wool.
2. Weigh the empty oil flask and fill it with the petroleum ether (say 50-100 ml). Place the oil flask over the heating mantle. The amount of solvent taken is about 1.5 times the capacity of the Soxhlet extractor.
3. Introduce the extraction thimble with sample into Soxhlet extractor (middle piece) over a pad of cotton wool so that the top of thimble should be well above the siphon end of extractor. Place the extractor along with the oil flask and connect it with the condenser.
4. Start the cold water circulation in condenser. Switch on the heating mantle or water bath. As soon as the ether begins to boil adjust the heat to 40-60°C or 60-80°C as per the boiling point of petroleum ether used.
5. Extract the sample for 4 hours (5 to 6 drops per second condensation) to 16 hours (2 to 3 drops per second condensation). In most of the cases 8 hours extraction is sufficient (10 cycles / hour).
6. After the period of extraction, remove the thimble from extractor and recover through the extractor all the ether form the oil flask leaving only few drops of ether in it.
7. Evaporate the residual ether of oil flask on water bath and place the oil flask in hot air oven at 80-100°C for two hours, cool it in dessicator and weigh it.

8. Continue cooling and weighing of the oil flask until the difference between two successive weighing is less than one milligram. Never leave more amount of petroleum ether in the oil flask and don't keep the flask in hot air oven otherwise fire may take place. If necessary, reserve the extracted sample in the thimble for the determination of crude fibre content.

Ether extract % (on as such basis) = 100 X Weight of ether extract / Weight of sample

= 100 X (B -A) / W

Where,

A = Weight of empty oil flask
B = Weight of oil flask + Ether extract
W = Weight of sample

Ether extract % (on DM basis) = 100 X Weight of Crude fat / Weight of sample on DM basis.

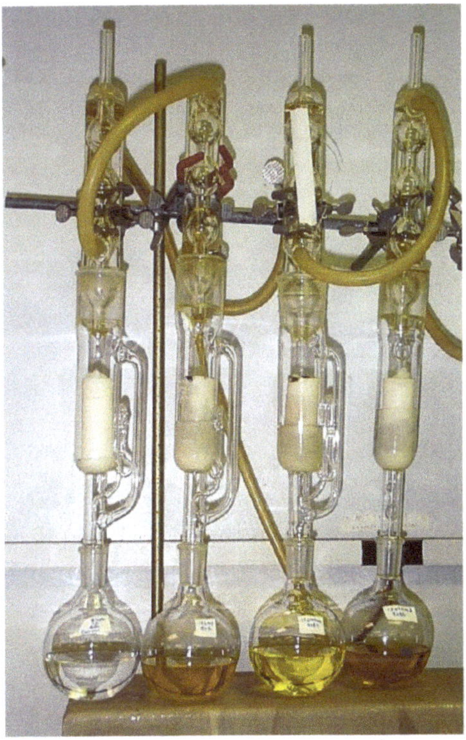

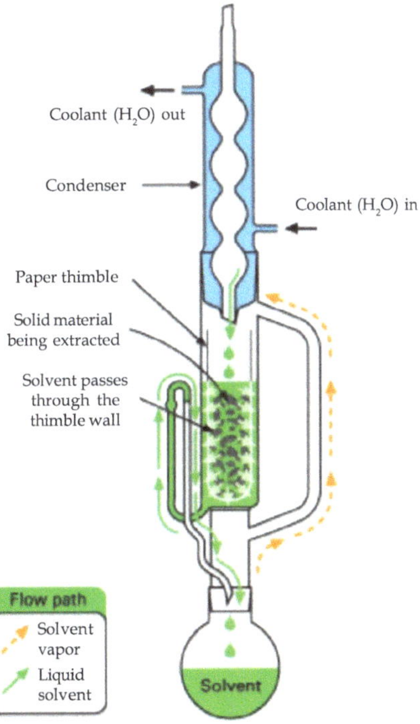

Soxhlet extraction apparatus

Exercise No. 9

Estimation of Crude Fiber (CF)

Principle

When a sample of moisture and fat free feedstuff is boiled with dilute acid and then with dilute alkali, the soluble carbohydrates and proteins go into the solution and are extracted. The undissolved residue left behind represents cellulose, hemicellulose, lignin and mineral matter. The cellulose, hemicellulose and lignin put together are called crude fibre (CF). When the residue after drying and weighing is ignited, the fibre being organic in nature is burnt whereas the ash or mineral matter is left over. The fibre is obtained by deducting the weight of ash from the weight of the dried residue. In simple way it can also be explained that the crude fibre is determined gravimetrically as part of the substance which is left when protein, soluble carbohydrate, fat, ash and water have been removed.

In feeding of simple stomached animals, the analysis of crude fibre content is very essential because the amount of crude fibre predicts the digestibility of the feedstuff. Higher the fibre content lesser will be the digestibility.

Equipment/glass wares

1. Analytical balance
2. Hot air oven

3. Hot plate heating mantle
4. Muffle furnace
5. Beaker (1000 ml), tall form spout less
6. Buchner funnel
7. Conical flask
8. Dessicator
9. Horn spatula
10. Measuring cylinder
11. Muslin cloth with 18 threads per centimetre square
12. Silica crucibles, large size
13. Suction flask with gasket
14. Suction pump
15. Wash bottle

Reagents

1. Acetone
2. Sulphuric acid (1.25 per cent, W/V)
3. Sodium hydroxide (1.25 per cent)

Procedure

1. Weigh accurately about 2g of dried feed material and extract the crude fat (ether extract) for about 8hrs with petroleum ether using soxhlet extraction apparatus.
2. Transfer the extracted feed material into tall spout less beaker (500/1000ml); mark the beaker for 200 ml with a marker.
3. Add 200ml of H_2SO_4 (1.25%) to the beaker and start the heater. In order to keep the volume constant in the tall form beaker or to check the evaporation, keep one flask filled with cold water over the beaker. Bulb condenser fitted over the beaker can also be used in lieu of round bottom flask.
4. When boiling starts, slow down the temperature of heater to let the contents only reflux for exact 30 minutes and then remove the beaker from the hot plate, filter the content from muslin cloth spread over the buchner funnel fitted with suction pump and

suction flask. Wash the residue with hot water for several times till the filtrate becomes acid free.

5. Transfer the residue (acid-free) by horn spatula from Buchner funnel to the same spout less beaker. Add 200 ml (1.25%) of NaOH solution and put again on the heater and switch on the heater and wait for boiling start and when boiling starts reduce the temperature to refluxing and keep refluxing for next half an hour.

6. Remove the beaker form the hot plate and filter the content again through the same muslin cloth on filter flask connected to a filter pump. Wash the residue with hot water for several times till free from alkali. Wash the residue with acetone.

7. Transfer the residue to a clean dry silica crucible. Dry the crucible and contents at 100°C ± 2°C in hot air oven for complete dryness. Cool the crucible in a dessicator and weigh the crucible containing dried residue.

8. After getting the sample smokeless on heater Incinerate the contents of the crucible at 550 - 600°C for two hours in a muffle furnace until all the carbonaceous matter is burnt.

9. Remove the crucible from furnace, cool in a dessicator and weigh the crucible containing ash. The loss in weight on ignition is due to the amount of crude fiber present therein.

Crude fiber assembly (Heater, spout less beaker, condenser)

Calculation

Crude fiber (%) on as such basis

= 100 X Weight of crude fiber / Weight of sample

= 100 X (A – B) / W

Where,

W = Weight of sample

A = Weight of crucible + crude fiber before ashing

B = Weight of crucible + ash

Crude fiber % (on DM basis)

= Weight of Crude fiber X 100/Weight of sample on dry matter basis

Exercise No. 10

Estimation of Nitrogen Free Extractive (NFE)

Principle

Carbohydrates are divided into two groups *i.e.,* crude fibre (CF) and nitrogen free extractives (NFE). The NFE comprises the sugars, starch and a large part of the material classed as hemicellulose.

The NFE is not determined by analytical procedures. This is calculated as the figure obtained when the sum of water, ash, protein, fat and crude fibre of a feed is subtracted from 100. Since the figure is determined by difference, it includes the cumulative errors of the other determinations and thus it is not an exact value.

Requirement

Following analysed volumes of a feedstuff are very much essential to derive the nitrogen free extractives of the same feedstuff.

Calculation

The nitrogen free extractives can be derived from the following formula:

NFE (%) = 100 - (CP (%) + EE (%) + CF (%) + Total Ash (%)

If NFE is to be expressed on as fed basis the following formula would be applicable:

NFE% = 100 - (Moisture % + CP% + EE% + CF% + Total Ash (%)

Exercise No. 11

Estimation of Calcium

Principle

When a solution containing calcium is treated with ammonium oxalate, all the calcium present is precipitated as calcium oxalate. The precipitate when dissolves in sulphuric acid, forming calcium sulphate and liberating free oxalic acid which is quantitatively estimated by titration against standard N/10 $KMnO_4$ to arrive at the calcium content present in the given solution. Following chemical reactions takes place in the estimation of calcium:

$$CaCO_3, CaO, Ca(OH)_2, CaSO_4 + HCl \longrightarrow CaCl_2 + H_2SO_4 + H_2O + CO_2$$

$$CaCl_2 + (COONH_4)_2 \longrightarrow (COO)_2\,Ca + 2NH_4Cl$$

$$(COO)_2\,Ca + H_2SO_4 \longrightarrow (COOH)_2 + CaSO_4$$

$$2KMnO_4 + 3H_2SO_4 \longrightarrow K_2SO_4 + 2MnSO_4 + 3\,H_2O + 5O$$

$$(COOH)_2 + O \longrightarrow CO_2 + H_2O$$

Equipment/glass wares

1. Beaker (250 ml)
2. Burette
3. Funnel and stand
4. Glass rod

5. Measuring cylinder
6. Pipette (25 ml or 50 ml)
7. Wash bottle
8. Whatman filter paper (No.40)

Reagents

1. Ammonia solution (1:4)
2. Ammonium oxalate solution (saturated)
3. Methyl red indicator
4. Potassium permanganate solution (N/10)
5. Sulphuric acid (dilute, 1: 4)
6. Hydrochloric acid (1%)

Procedure

The method of calcium estimation in feedstuffs is known as Talapatra method. This comprises three steps i. e. precipitation, filtration and titration.

Precipitation

1. Pipette out an aliquot (10 ml) of mineral extract into a beaker and add 2-3 drops of methyl red indicator solution.
2. Heat to boiling, cool the contents and add 10ml of saturated ammonium oxalate solution slowly drop by drop with constant stirring until the precipitate is coarsely granular.
3. Heat the contents till boiling, cool and add NH_4OH (1:4) drop by drop till the colour develops faint pink.
4. If the red colour of the solution changes to orange or yellow, add dilute hydrochloric acid drop wise until the colour again changes to pink. Here one notable point is that the adjustment of pH of the solution is very important as otherwise magnesium will also get precipitated above this pH range in an alkaline medium.
5. Leave overnight to allow the precipitate to settle.

Filtration

Filter the supernatant liquid through ash free filter paper (Whatman No.40). Wash the precipitate with small quantities of hot water several times till free from oxalate. To check whether the precipitate is oxalate free, collect half a test tube of the filtrate, acidify it with a few drops of concentrated sulphuric acid, add one drop of N/10 potassium permanganate solution and heat the solution. Pink colour of permanganate remains as such indicates the absence of oxalate.

After making the precipitate oxalate free, transfer the filter paper with precipitate carefully back to the same beaker. Unfold the filter paper along the side of the beaker and wash the precipitate down with a jet of hot distilled water (about 50 ml) in the beaker. Wash the precipitate with sufficient quantity of H_2SO_4 (20 per cent, 10 ml) till the precipitate completely dissolves. Add conc. H_2SO_4, if necessary.

Titration

Heat the contents in the beaker to 60-70°C (when first bubble comes out) and titrate with N/10 $KMnO_4$ solution to a faint pink end point and the colour must persist for more than 30 seconds. Record the volume of N/10 $KMnO_4$ solution for the titration.

Calculation

1000 ml of 1N $KMnO_4$ = 20 g calcium
1 ml of 1 N $KMnO_4$ = 0.02 g calcium
1 ml of N/10 $KMnO_4$ = 0.002 g calcium

Calcium (%) on as such basis
= 0.02 X ml of std $KMnO_4$ used X N X V X 100/W X A

Where,
N = Normality of the standard $KMnO_4$ solution
V = Volume in ml of the mineral extract made
W = Weight of sample
A = Aliquot of the mineral extract used for determination in ml

Calcium (%) on dry matter basis
= Weight of Calcium X 100/Weight of sample on dry matter basis

If the value of calcium (Ca) is to be expressed in terms of calcium oxide (CaO) or vice versa the following conversion equation may be applied:

CaO = Ca X 1.4
Ca = CaO X 0.714

Exercise No. 12

Estimation of Phosphorus

Principle

Phosphorus is the chief acid forming mineral constituent of plant and feeds. Phosphorus present in the ash solution (Mineral extract) is precipitated as ammonium phosphomolybedate. The precipitate is washed with dilute potassium nitrate till free from acid and dissolved in a known excess of standard sodium hydroxide solution. The excess alkali is back titrated against standard acid to arrive at the exact quantity of standard solution required.

Following chemical reaction takes place in phosphorus estimation

$$(NH_4)_3PO_3\ 12MoO_3 + 23\ NaOH \longrightarrow 11\ Na_2\ MoO_4 + (NH_4)_2\ MoO_4 + Na\ NH_4\ HPO_4 + 11\ H_2O$$

Equipment/glass wares

1. Mantle heater
2. Beaker (150 or 250 ml)
3. Burette and stand
4. Conical flask
5. Funnel and stand
6. Glass rod
7. Measuring cylinder

8. Pipette (5, 10, 25, 50 ml)
9. Tongs
10. Whatman filter paper (No.42)

Reagents

1. Ammonium molybdate solution: Take 200 g of powdered ammonium molybdate $(NH_4)\ 6Mo_7O_{24}.4H_2O)$ in a stoppered graduated cylinder of 1000ml capacity, add to it 800 ml of distilled water and shake well for 25 minutes to dissolve the ammonium molybdate. Add gradually 25 % (w/v) ammonium hydroxide solution till the solution is clear (about 100 to 140 ml of ammonium hydroxide may be required). Avoid adding excess of ammonia. Make up the volume to one litre. If necessary, filter the solution.
2. Nitric acid solution (1 to 3%)
3. Phenolphthalein indicator (Dissolve 0.1 g of phenolphthalein in 100 ml of 60 percent rectified spirit)
4. Potassium nitrate solution (2 per cent; w/v)
5. Standard acid solution (N/7or N/10)
6. Standard sodium hydroxide solution (N/7or N/10)

Procedure

The estimation of total phosphorus can be completed in three different stages viz, Precipitation, filtration and titration.

Precipitation

1. Pipette out an aliquot (25 or 50 ml) of the mineral extract (prepared earlier) in a clean and dry beaker (150 or 250 ml capacity).
2. Take 10 ml of conc. HNO_3 in a test tube and 10 ml of ammonium molybdate reagent into another test tube and add them simultaneously into the beaker containing pipetted aliquot. Stir the contents with the help of a glass rod till the precipitate appears. Allow the precipitate to settle down overnight.

It is to be noted that the temperature developed in the molybdate solution is sufficient to precipitate all the phosphorus present in the aliquot. Under no circumstances the phosphomolybdate precipitate should be heated either on a water bath or directly over a burner so as to avoid precipitation of molybdic anhydride.

Filtration

1. Filter the precipitate through a filter paper (No.42) over an ordinary funnel. As far as possible only the supernatant liquid is passed through the filter paper retaining the precipitate in the beaker.
2. When the supernatant liquid is decanted off, wash the precipitate twice with dilute HNO_3 (2 per cent) and then with potassium nitrate solution (2-3%, w/v) until the washings are free from acid. If ordinary funnel and filter paper are used, test the filtrate for acidity with litmus paper or by collecting the filtrate in a test tube, adding a few drops of phenolphthalein indicator solution and one drop of the standard NaOH solution. If the pink colour appears with one drop of standard alkali the precipitate is free from acid.

Note that the ammonium phosphomolybdate precipitate is sparingly soluble in water and hence should be washed with 2 per cent HNO_3 solution and then with 2 percent potassium nitrate solution but not with water.

Titration

1. Transfer the precipitate to the original beaker in which the precipitation was carried out. Wash the funnel with a small amount of distilled water.
2. Dissolve the yellow precipitate by adding sufficient quantity of standard NaOH solution from a burette. Add 5ml of standard NaOH solution in excess. See that no yellow precipitate sticks to the filter paper. Note the total volume of the standard NaOH solution added.
3. Add about 10 drops of phenolphthalein indicator to the beaker containing the dissolved precipitate and titrate the solution with the standard acid solution (HNO_3, Hcl or H_2SO_4). Note the volume of acid consumed in titration.

Calculation

1000 ml of 1 N NaOH = 1.347 g of phosphorus

1 ml of 1 N NaOH = 0.001347 g of phosphorus

1 ml of N/10 NaOH = 0.0001347 g or 0.1347 mg of phosphorus

1 ml of 0.14 N (N/7) NaOH = 0.0001924 g or 0.1924 mg of phosphorus

Phosphorus (%) on as such basis
= (X-Y) X 0.001347 X N X V X 100 / W X A

Where,

X = Volume in ml of the standard NaOH solution used in dissolving the precipitate

Y = Volume in ml of the standard acid used in titration to neutralise the alkali

N = Normality of the standard NaOH solution

V = Volume in ml of the mineral extract made after ashing

W = Weight of the sample

A = Aliquot in ml of the mineral extract pipetted out for precipitation

Phosphorus (%) on DM basis
= Weight of P X 100 / Weight of sample taken on dry matter basis

If the phosphorus (P) volume is to be expressed in phosphorus pentaoxide (P_2O_5) or vice versa, the following conversion factor may be applied:

$P_2O_5 = P \times 2.29$

$P = P_2O_5 \times 0.436$

Exercise No. 13

Detergent Method of Forage Analysis

To overcome the limitation of proximate analysis P.J. Van Soest in 1963 developed a method which makes use of the concept that the dry matter of plant origin consists of two principal parts:

1. Cell contents
2. Cell wall contents

The method is efficient to take care of the defects in the principle of estimating crude fiber and NFE by proximate analysis. Van Soest has partitioned the carbohydrates into various fractions by a system of analysis using detergents. Cell contents are soluble in neutral detergent includes lipids, sugars, organic acids, starch, soluble proteins, NPN compounds and other water soluble materials, where as cell wall (fiber) is insoluble in neutral detergent but it is soluble in acid detergent includes hemicelluloses, cellulose, lignin, lignified nitrogen compounds, heat damages proteins, keratins and silica. This method of partitioning carbohydrate is essential to be expressed in case of ruminants, due to their capacities for making use of different fractions of crude fibre especially the cellulose and hemicellulose, while the simple stomached animals can rarely do so.

13.1 Estimation of cell contents and cell wall constituents

Principle

The neutral-detergent procedure for cell wall constituents is a rapid method for analysing the total fiber in vegetable feedstuffs. It appears to divide the dry matter of feeds very near the point that separates the nutritionally available (98%) and soluble constituents from those that are incompletely available and dependent on a microbial fermentation. The cell contents are determined as the difference between the percent cell wall and 100 percent.

Apparatus

1. Refluxing apparatus:
   (a) Tall pyrex or corning beakers (spout less) of about 500 ml capacity.
   (b) Round bottom flask as condenser.
2. Sintered glass crucibles with coarse porosity (Grade I) of about 50 ml capacity.
3. Electronic balance
4. Vacuum pump
5. Hot plate
6. Wash bottle
7. Hot air oven and muffle furnace

Reagents

Neutral-detergent solution

Distilled water: 1/litre

Sodium lauryl sulphate: 30 g

Disodium ethylene diamino tetra acetate (EDTA) dehydrate: 18.61g

Sodium borate decahydrate ($Na_2B_4O_7.10H_2O$): 6.81 g

Disodium hydrogen phosphate, (anhydrous): 4.56 g

2-ethoxyethanol (ethylene glycol monoethyl ether):10 ml

Put EDTA and $Na_2B_4O_7.10H_2O$ together in a large beaker, add some of the distilled water and heat until dissolved, then add to solution containing disodium lauryl sulphate and 2-ethoxy ethanol. Put

Na_2HPO_4 in a beaker, add some of the distilled water and heat until dissolved, then add to solution containing other ingredients. Check pH range 6.9 to 7.1. If solution is properly made pH adjustment is rarely required.

Acetone

Use grade that is free from color and which leaves no residues upon evaporation.

Procedure

1. Take 0.5 to 1.0 g air dry sample ground (1 mm) into a beaker of the refluxing apparatus.
2. Add 100 ml. (preheated) neutral detergent solution, and reflux for 60 minutes, time starting from the onset of boiling.
3. Filter off the regent in a per-weighed empty sintered glass crucible, wash thrice with hot distilled water under vacuum, remove vacuum, break up mat and wash crucible with hot water.
4. Wash twice with acetone in same manner and suck dry. Dry crucible at 100°C for 8 hrs or overnight and weigh it.
5. Report yield of recovered NDF as percent of cell wall constituents, estimate cell soluble material by subtracting this value from 100.
6. Ash residues in the crucible for 3 hrs at 500-550°C and weigh. Report ash content as ash insoluble in neutral-detergent.

Calculation

Cell wall constituents (%) (NDF) = (weight of crucible + cell wall constituents) - weight of crucible X 100/Weight of sample on dry matter basis

Cell contents (%) = 100 - cell wall constituents.

13.2 Estimation of Acid detergent Fiber (ADF)

Principle

The acid-detergent fiber procedure provides a rapid method for lignocellulose determination in feeds stuffs. The residue also includes silica. The difference between cell walls and ADF is an estimate of hemicellulose, however, this difference does include some protein attached to cell walls. The acid detergent fiber is used as a preparatory step for lignin determination.

Apparatus

Same as used in NDF estimation

Reagents

a. Sulphuric acid (H_2SO_4)
Regent grade, standardized to 1N.

b. Cetyl trimethyl ammonium bromide (CTAB) Technical grade: Add 20 g CTAB in the 1N H_2SO_4 solution and Stir.

c. Acetone -Use grade that is free from color and leaves no residue upon evaporation.

Procedure

1. Weigh 1 g air dry sample ground to pass 1 mm. screen or approximate equivalent of wet material in a beaker suitable for refluxing.
2. Add 100ml acid-detergent solution. Heat to boiling in 5 to 10 minutes. Reduce heat as boiling begins, to avoid foaming. Reflux 60 minutes from onset of boiling; adjust boiling to a slow, even level.
3. Filter on a previously weighed sintered glass crucible. Wash with hot distilled water 3-4 times breaking the mat. Repeat wash with acetone twice or until it removes no more color and suck dry.
4. Keep it for drying in oven at 100°C for 8 hrs or overnight and weight after cooling of crucible in desiccator.

Calculation

Acid detergent fiber % (on dry matter basis) = (weight of crucible + fiber) - empty weight of crucible X 100/ Weight of sample on dry matter basis.

13.3 Estimation of Acid detergent lignin (ADL)

Principle

In the acid-detergent lignin procedure, the acid detergent fibre (ADF) procedure is used as a preparatory step. The detergent removes the protein and other acid-soluble material that would interfere with the lignin determination. The ADF residue consists of cellulose, lignin and acid-insoluble ash (mainly silica), treatment with 72 percent sulphuric acid dissolves cellulose. Ashing of the residue will determine the lignin fraction.

Apparatus

1. Same as required for NDF estimation
2. Glass tray
3. Hot air oven and Muffle furnace
4. Glass rod
5. Dessicator

Reagents

H_2SO_4, 72% by weight or 58.3 % (v/v)

Take 417 ml water in a volumetric flask and add 583 ml pure H_2SO_4 slowly with occasional swirling. The flask must be cooled in water bath (Sink) in order to add the required weight of acid.

Procedure

1. Prepare the acid detergent fiber.
2. Place the crucible in the glass tray. Have one end of the tray 2 cm. higher so acid will drain away from the crucibles.
3. Cover the contents of crucible with cooled (15°C) 72 percent H_2SO_4 and stir with a glass rod to a smooth paste, breaking all lumps. Fill crucible about half way with acid and stir. Let glass rod remain in crucible, refill with 72 percent H_2SO_4 and stir at hourly intervals as acid drain away. Crucible does not need to be kept full at all times. Three additions suffice. Keep crucible at 20 to 23°C. After 3 hours, filter off as much acid as possible with vacuum, and then wash contents, with hot water until free from acid. Rinse and remove stirring rod.
4. Dry crucible overnight at 100°C and weigh
5. Ignite crucible in a muffle furnace at 500 to 550°C for 3 hours, and then cool and weigh.

Calculation

Acid detergent lignin (ADL) % = Weight of crucible and lignin - Weight of crucible and ash X 100 / Weight of sample on dry matter basis.

Exercise No. 14

Quantitative Detections of Undesirable Constituents and Common Adulterants of the Feed

The undesirable constituents includes the toxicants and other adulterants in the feed which are not suppose to be there in the feed and presence of these substances either make the nutrients of feed less utilizable or even the feed become toxic for consumption by animals. Hence it is very essential to find out the presence and also the amount of that substances on which basis decision may be made that the particular feed stuff or ingredient is safe to be consume by the livestock or that has to be rejected. The common undesirable constituents of the feeds are given below in detail:

14.1 Determination of tannins

Tannins are poly-phenolic compounds divided into 2 main groups-hydrolyzable and condensed. Hydrolyzable tannins contain a polyhydric alcohol usually glucose esterified with gallic acid or with hexahydroxydiphenic acid. Condensed tannins are mostly flavanoids and are probably polymers of flavan-3-01 (catechin) and these can not be hydrolyzed to simple components. Due to presence of tannins in feeds, there is decrease in feed consumption and digestibility particularly in simple stomach animals.

Principle

Tannins can be estimated as quercitannic acid by titrating against potassium permanganate in the presence of indigo carmine.

Equipments/glass wares

1. Hot plate
2. Volumetric flasks (250, 500 ml)
3. Amber coloured bottle
4. Round bottom flask
5. Burette
6. Conical flasks (100 ml)
7. Funnel
8. Spout less beaker
9. Pipettes

Reagents

1. Oxalic acid (0.1 N): Dissolve 1.575 g oxalic acid in 250 ml distilled water.
2. $KMnO_4$ solution (0.1 N): Dissolve 1.34 g $KMnO_4$ in 1 litre distilled water. Store in amber coloured bottle overnight and standardise against 0.1N oxalic acid next day.
3. Indigo carmine solution: Dissolve 6 g indigo carmine (sodium indigotin disulphonate) in 500 ml distilled water by heating. Cool and add 50 ml dilute sulphuric acid (20%) and make up the volume to 1000 ml with distilled water.

Procedure

1. Take 5g dried, powdered and fat free sample and reflux it in 300 ml distilled water for 1 hour. Collect the filtrate by decantation.
2. Reflux the residue with another 200 ml distilled water again for 15 minutes. Collect the filtrate and cool.
3. Make the volume to 500 ml and filter through filter paper (Whatman No.1).
4. Take 101 ml of distilled water and add 2 ml indigo carmine solution. Titrate it against 0.1 N $KMnO_4$ solution and record the volume of $KMnO_4$ used (Vo).
5. Take 1ml aliquot (extract) and add 100 ml distilled water and 2 ml indigo carmine solution. Titrate in against 0.1 N $KMnO_4$ solution and record the volume of $KMnO_4$ used (V).

Calculation

Tannic acid in sample (DM basis) = (V-Vo) X 0.006235 X 500 / W

Where, V = Volume of 0.1 N $KMnO_4$ used in case of sample

Vo= Volume of 0.1 N $KMnO_4$ used for blank

W = Dried weight of sample refluxed

The factor 0.006235 has been used as 1ml 0.1N $KMnO_4$ = 0.006235 g tannic acid.

14.2 Determination of Trypsin inhibitor

Principle

The trypsin inhibitor activity is measured indirectly by inhibiting the activity of trypsin. A synthetic substrate (BAPNA) is subjected to hydrolysis by trypsin to produce yellow coloured p-nitroanilide. The degree of inhibition by the extract of the yellow colour production is measured at 410 nm.

Reagents

1. Glacial acetic acid (30%, v/v)
2. *Trypsin*: Dissolve 6.25mg of lyophilized trypsin and make up to 25 ml with 0.001M HCl. Dilute 2ml of this solution to 25 ml for assay.
3. *Substrate*: Benzoyl-DL-arginine-paranitroanilide (BAPNA). Dissolve 40 mg BAPNA in 0.5 ml of dimethyl sulphoxide and then make up to 100 ml with Tris-HCl buffer (pH 8.2).
4. *Tris-HCl buffer (pH 8.2):* Weigh 6.05g Tris-hydroxymethyl aminomethane and 2.94g of $CaCl_2.H_2O$, dissolve in 900ml water, adjust to pH 8.2 with dilute HCl and make up to 1000 ml with distilled water.
5. *Source of trypsin inhibitor (TI):* Extract 0.5g sample in 25 ml water by grinding in a pre chilled mortar and pestle. Extract the ground sample in a refrigerator for 2-3h with occasional shaking for complete extraction of TI. Centrifuge the homogenate at 12,000 rpm for 20 min at 4-6°C. Dilute 1ml of the supernatant to 10 ml with distilled water and use as TI source.

Procedure

1. Pipette out 0 and 1 ml of extract in duplicate sets of test tubes, one to serve as endogenous (E) and the other test (T).
2. Make up the volume to 2 ml with buffer in the endogenous set.
3. Make up the volume to 1 ml in the test set.
4. Add 1ml of trypsin solution (20μg) to each tube in the test set. Pipette out into a separate test tube 1ml of buffer and 1ml of trypsin solution for standard (S).
5. Incubate all the tubes in a water bath at 37°C.
6. After a few minutes, add 2.5 ml of substrate (1mg BAPNA) to each tube.
7. Allow the reaction to proceed for 10-60 min at 37°C.
8. Stop the reaction by adding 0.5 ml of 30% glacial acetic acid.
9. Read the absorbance at 410 nm in a spectrophotometer.
10. Determine the protein content in the extract by Lowry's Method.

Calculation

Find out T-S absorbance.

Plot the absorbance against the volume of extract.

Determine the aliquot size of the extract required to inhibit 50 per cent of the trypsin activity (S/2). That aliquot size is considered to be one unit of trypsin inhibitor.

One unit of activity corresponds to that amount of trypsin inhibitor in μg protein which gives 50 per cent inhibition of enzyme activity under experimental conditions. The trypsin inhibitor activity is expressed as trypsin inhibitor units (TIU) per gram sample or per mg protein (the dilutions of trypsin inhibitor source were made in such a way that 0.5 ml produces 50 per cent inhibition).

14.3 Estimation of saponins

Reagents

1. Methyl alcohol
2. n-Butyl alcohol
3. Solvent ether (Redistilled)

Method

Extract 5 g of finely ground feed sample for 10 hours in a soxhlet apparatus with methyl alcohol. Concentrate methyl alcohol extract (to about 10ml) on heating mantle and later it is diluted with 20ml distilled water. Transfer it to a separating funnel (250ml). To ensure quantitative transfer three washings (5, 3 and 2 ml of distilled water) should be given and add to separating funnel. Add 10ml n-Butyl alcohol and shake for a minute. The layer containing n-Butyl alcohol extract, is collected in a 50 ml beaker and repeat the same process twice using 5ml portions of n-Butyl alcohol. Concentrate the n-Butyl alcohol soluble fraction under reduced pressure. Dissolve it in 5ml of methyl alcohol drop by drop in three volumes of solvent ether while stirring with the help of glass rod. A white precipitate of saponins will be obtained. Filter it through pre-weighed whatman filter paper no 1. Filtration should be done immediately as otherwise the precipitate will stick on the wall of beaker on keeping. The beaker containing solution of saponins should be washed thrice with methyl alcohol (1ml portion) to ensure quantitative precipitation. The beaker containing the precipitate should also be washed thrice with solvent ether (1 ml portion) Dry the filter paper having saponins precipitate under vacuum and weigh. The weight of saponins is expressed on the basis of dry weight of the sample taken.

Calculation

Total saponin (%) = (W2- W1) X 100 / W

Where,

W = Dried weight of sample

W1 = Dried weight of filter paper

W2 = Dried weight of filter paper along with precipitate of saponins

14.4 Estimation of mimosine

Chemicals and reagents

1. N/10 Hydrochloric acid
2. Activated charcoal
3. 0.5 % Ferric chloride solution
4. Standard mimosine solution (1 mg/ml in N/10 Hcl)

Procedure

1. Extraction

1. Take 1g dried and ground sample in a beaker. Add 30 -40ml N/10 HCl, stir with glass rod and boil gently for 5 minutes on hot plate.
2. Cool and allow the material to settle down
3. Filter through a nylon cloth in a 100 ml measuring cylinder.
4. Again boil the material (residue) with 20-25 ml N/10 Hcl, cool and filter for two more times
5. Make up the volume of filtrate to 100ml with N/10 Hcl.

2. Decolourisation

Take 10 ml filtrate in a conical flask and add 1.0ml of N/10 Hcl and a pinch of activated charcoal. Shake to mix thoroughly and heat for 5 minutes. Cool, allow to settle and filter through whatmen filter paper no 1. If the filtrate is not yet clear, again treat it with charcoal and collect the clear filtrate.

3. Estimation

Take 1ml filtrate and add 8.6ml N/10 Hcl and 0.4ml of ferric chloride solution and read it at 535nm. Also run a blank with it. Compare the O.D. of sample - blank with standard curve.

Preparation of standard curve

Take 0, 0.1, 0.2, 0.4, 0.6, 0.8 and 1.0 ml of standard solution and proceed as follows:

Tube no. in duplicate blank	$Fecl_3$ solution solution	Standard mimosine	Vol of N/10 Hcl
1	0.0	9.6	0.4
2	0.1	9.5	0.4
3	0.2	9.4	0.4
4	0.4	9.2	0.4
5	0.6	9.0	0.4
6	0.8	8.8	0.4
7	1.0	8.6	0.4

Take O.D. at 535nm for the sample and draw graph against the concentration of mimosine.

14.5 Methods of quality control for feeds

The concept of quality in feed stuffs generally refers to the selection of proper type of ingredients required for achieving the desired specifications which make them suitable to be consumed by animals without any negative effect.

Quality control refers to the cycle of activities for achieving and maintaining the intended objectives or standards. The standards specify the attributes or characteristics, which can be used for measuring or identifying quality control. The objective of quality control of feed stuffs is to ensure that a consumer obtains feeds that are unadulterated, true to their nature and give desired results. It may mean that feed should be of the acceptable quality.

The evaluation of quality of the feed stuffs is based on the evaluation of the feed ingredients that will be used for formulation of ration. To control quality of feeds it is necessary to have a reliable method of quantitatively appraising specific characteristics of the feed. The essential characteristics of the feed, such as moisture, fat content, protein and crude fiber are required to be compared with the stipulated requirements. For getting perfect quality control following observations usually made:

1. Visual evaluation

Among the organoleptic test, visual examination is the best. An experienced person can judge the quality of feeds specially grains, brans, oil cakes and fish with fair degree of accuracy by visual examination.

(a) Examine the feedstuffs for natural colours, consistency, presence of foreign materials, mould growth, cake and clump formation, weevil attack and any other abnormalities. Preparing feed with such spoiled feedstuffs or caked up mouldy feed results in poor production and high mortality.

(b) Mouldy grain will have greenish, greyish or blackish discoloration. Prolonged storing or storing with high moisture level leads to clump and cake formation as well as mould growth and burning appearance.

(c) Size of grain is an indicator of energy value, the smaller the grain, the lower the ME value because of increased proportion of seed coat. Weight of a fixed number of grains can be used to evaluate the cereals (weight of 100 on a unit volume of grains). This technique is called test weight.

(d) Adulteration of ground nut oil Cake with rice bran or any cheaper oil cakes and fish with prawn heads, crabs, squilla, shells etc. can be detected by careful visual examination as well as by feed microscopy. White fish is preferred to dark fish due to better keeping quality of the former because of low oil content.

(e) Adulteration of oil seed meals with fibrous materials will lower the nutritive value and can be easily detected. In rice polish, contamination with husk can be visually identified. Clumps or crystals in mineral ingredients are not suitable for premixing.

Identification of Feedstuffs Based on Colour

Deviation from natural colour indicates the adulteration and damage

Colour of particles	Probable ingredient
White	Limestone, milk powder, dicalcium phosphate, other mineral salts, starch
Buff	Soybean meal, bone meal, dried whey, dolomite, bran, husk, rock phosphate
Black	Charcoal, charred bone, blood meal, sesame cake, sunflower, niger cake, kapok seeds, leather meal, meat meal
Brown or reddish	Blood meal, meat meal, maize bran, husk, hulls
Grey	Dicalcium phosphate, rock phosphate, hair meal, leather meal, meat and bone meal
Yellow or orange	Yellow maize particles, dehulled cotton seed meal, soya hulls

2. Evaluation by touching

Particle size: fine, medium and course. Any deviation from normal proportions of these particle sizes indicates adulteration and contamination with damaged ingredients.

(a) Put your hand deep into a bag of grains. A temperature differential between inside and outside of the bag is not felt if grains are properly dried. If the moisture level is high, the grains inside the bag will be cooler during winter and hotter during summer than the outside temperature.

(b) Place about one spoon of rice polishing, rice bran or deoiled rice bran in left hand palm. Take a pinch of it and rub between fingers. If it is too coarse and rough it may be concluded that the bran is adulterated with rice husk.

(c) Press a handful of rice polishings in your fist and gently release the fingers. A good quality rice polish will show the finger impressions whereas the one adulterated will crumble indicating of fibre level of more than 12 per cent. If poultry mash shows such finger impression, it is indicative of high moisture level and the feed with normal moisture will break easily.

(d) In broken rice containing bran, if a significant amount of bran particles adhere to the palm, it is oil containing polish and the ME value will be higher but they cannot be stored for a longer period. If only a few particles are adhering, the ME value will be lower.

(e) Fish with more moisture level will be hot, heavy dark, more flexible and moist to touch.

(f) Clumps can be formed due to high moisture content or improper storage or due to packing immediately fresh warm solvent extracted meal. The clumps formed due to packing of warm material crumble on application of light pressure while those formed due to excess of moisture are hard.

3. Evaluation by taste

(a) Each ingredient has a different taste, any change in taste can be detected. Fresh feeds will have a desirable taste. Oil or oil in feeds develops rancidity leading to undesirable burning taste due to formation of free fatty acids. Oil feeds will have undesirable stale and musty taste.

(b) Bitterness in grains, soybean meal, sunflower oil meal and groundnut meal indicates presence of mycotoxins.

(c) By biting and tasting oilcakes one can find the freshness, rancidity, mould, mustiness and adulteration, if any. Presently, most of the oil cakes are adulterated with rice bran, castor seed, rubber seed and other cheaper oil seed cakes. Most of the adulterants can be detected by feed microscopy, chewing, tasting; colour and odour.

(d) Rice bran and polish adulterated with rice husk will have a throat burning taste with a feeling to spit the fibrous portion immediately.

(e) It is easier to determine moisture level in grains and oil seed cakes by biting and chewing them. A dry feed stuff will be crisp, hard and brittle to bite.

(f) Chewing fish meal, broken rice etc. adulterated with sand and pebbles will hurt the teeth making chewing difficult and force you to spit it out.

(g) By licking the fish and tasting a small sample of fish meal, level of salt can be found out. A salty taste comparable to that of a pickle is equivalent to about 5 per cent salt level. Salty taste comparable to that of regular dishes may have 2-3 per cent salt.

4. Evaluation by smell

(a) One can detect the rancidity developed due to prolonged or improper storage of rice polish, oil seeds, oily fish, meat meal, fat, oil or other oil/ fat rich feedstuffs by smelling. Use of rancid feeds causes destruction of fat soluble vitamins, decrease in feed intake and production.

(b) Sour odour is indicative of fermenting of mouldy grain or insect infestation. Musty odour indicates the presence of insects or fungal contamination. Odour similar to petroleum products is suggestive of excess of pesticides or fungicides.

(c) Meat meal adulterated with tannery waste, hair meal and silk worm pupae meal will have a strong leathery / tannery aroma or rancid animal fat smell.

(d) Heat damaged ingredients can be detected by their smell even when there is no visible change in colour.

(e) In good quality meat cum bone meal, the tallow smell is similar to that of ghee, if it is of good quality. Leathery smell in meat cum bone meal indicates adulteration with leather meal.

5. Audio evaluation

(a) Well dried grains and fish on shaking in hands will give you a dry metallic sound where as high moisture grains and fish will give soft dull sound.

(b) When the grain is bitten, a typical sound is heard if it is dry.

(c) Well dried feeds and feedstuffs are free flowing. Well dried fish can be ground easily and will not stick together.

6. Feed microscopy

Determining and evaluating the quality of feed ingredients and finished feeds and the extent of adulteration and contamination with the help of microscope is called as feed microscopy. This is an improvement over direct visual examination. Here the feedstuffs are examined under a wide field microscope.

Feed microscopy involves identification and evaluation of feed ingredients and foreign materials alone or in mixture either via surface features or cellular characteristics. If ingredients and contaminants are separated and their proportions measured, then feed microscopy can be used for quantitative evaluation of feed ingredients. Therefore, feed microscopy can satisfactorily tell the purity of feed ingredients and the acceptance and rejection decisions are made just after comparison with standard photomicrograph catalogue. Adulterants like husk, hulls, stones, foreign bodies, substitution with cheaper materials can be easily detected by microscopical examination. Presence of hair/leather meal in fish/meat meal, castor seed meal in sesame/ sunflower meal and similar cheaper substitutes can be detected.

7. Simple chemical methods for identification of feed supplements

S.No.	Reagent	Reaction	Compounds identified
1	2.5% ammonium molybdate solution	Effervescence and no precipitate	Carbonates, viz. calcium carbonate, sodium bicarbonate
2	Do	Yellow precipitate and no effervescence	Phosphates, viz. disodium phosphate, monosodium phosphate, bone meal.
3	Do	Effervescence & yellow precipitate	Dicalcium phosphate
4	1% silver nitrate solution	White precipitate	Salt
5	Do	No precipitate	Salt is absent
6	Distilled water	White solution	Milk products may be present
7	Do	No white colour	Milk products absent

8. Some physical methods

(a) The common contaminant or adulterant is sand. The percentage of sand can be estimated by mixing a weighed quantity of grain with water, separating the grain, draining the water and weighing the sand.

(b) To detect husk in a feedstuff, winnowing is the best method. Husk separated from a weighed quantity of the material allowed to fall against blowing light wind can be weighed.

(c) Sieving can be used to differentiate contaminants based on particle size.

(d) Test weight can be used as a measure of bulk density. As bulk density increases, nutrient content also increases. A litre measure of the ingredient is weighed to arrive at the test weight or bulk density, for example in maize, the normal test weight is 725-775 kg/m3 or 725 to 775 kg/100 litres or one litre weighs 0.725-0.755 kg. Oil cakes have to be ground before test weight. The test weight of the ingredients estimated can be compared with the standard.

9. Miscellaneous tests

(a) *Test for Level of Roasting in Soybean Meal*: Over roasting and under roasting are not desirable in case of soybean meal. A mixture containing cresol red + urea + thymol blue + glycerol is use for this rapid field test. Spread the sample uniformly in a thin layer on Petri dish or white glazed paper. Spray the above reagent over the particles. In case of under roasting, only about 10 per cent soya develops red colour that too slowly whereas there is no development of red colour in case of over roasting.

(b) *Test for Fish Meal*: The fish meal available in our country is an admixture of poor quality fish, squilla, prawn heads, crabs, shells and other sea waste. Previously urea was added to increase protein level in such fish meal. At present they are adding meat meal, leather meal, silk worm pupae, feather meal, hair meal etc which are rich in protein but low in digestibility. Use of such fish meal should be avoided. Besides that, fish meal also contains a lot of sand and salt. For detection, mix about 25 g of fish meal in about 100 ml of water in a glass tumbler. Stir well with a spoon. Taste a drop of this water. If the salty taste of it is comparable to that of soup; the fish meal will be having 4-5 percent salt. Beyond this it is not acceptable. Examine the level of sand settled at the bottom of glass. This gives a fairly good idea about the sand and silica level.

(c) *Tests for Rice Polish*: Hold sufficient quantity of rice polish in hand and lighten your fist. Release your fingers. The impressions of fingers will remain with crumbling in good quality rice polish. Also, rice polish must be fresh with fine aroma and taste, soft to touch and free from insect attack and rancidity.

(d) *Tests for Grains*: Visualize for their colour, consistency, insect attack and other foreign materials like sand, stones, cobs, husk, bran etc. Examine the germ portion of the grain for mould growth. Randomly examine 50-100 grains from different bags for mould, if any. If more than 2 percent grains show mould growth, it may be rejected. If 1-2 per cent of the grains are affected, send the sample to laboratory for mycotoxin estimation and confirmation.

By biting, touching and shaking, the moisture level can be assessed with fair degree of accuracy. If more precision is needed, weigh 100 g of ground grain to 1 g accuracy, heat in a frying pan with constant stirring for about 10min., do not char it. Cool and weigh again. This will give dry matter level.

(e) *Tests for Oil Seed Cakes*: The quality, freshness and adulteration level in oilcakes can be assessed with fair degree of accuracy by organoleptic tests. The moisture can be estimated as in case of grains. By seeing the colour, smelling and chewing one can determine the extent of roasting in case of soybean meal.

Exercise No. 15

Silage Making

Silage is the green material produced by controlled anaerobic fermentation of the green succulent fodder crop retaining its moisture content. The process of preparing silage is known as **Ensiling** and the pit in which silage is prepared is called **silo.**

Crops suitable for silage making

The fundamental behind the conservation of fodder as silage is the **acidic condition**, which is produced by the anaerobic fermentation of carbohydrates rich fodder. Thus the necessity of the fodder crop being used for silage preparation is the presence of high amount of fermentable carbohydrates and the natural carbohydrate rich crops are the cereal crops like maize, jowar or bajra and grasses or their combination in 3:1 with leguminous crops. Otherwise high moisture leguminous crop after wilting in combination of molasses may be used.

15.1 Laboratory ensiling of green fodders

Small quantities of silage can be prepared at laboratory with green fodders to study physical and chemical properties of silage and the changes that occurs during silage making. Under laboratory conditions silage can be prepared by using polythene bag silos.

Equipments

1. Hand driven chaff cutter
2. Containers *i.e.* thick polythene covers, plastic bags, plastic tins, concrete bins etc.

Steps in ensiling

1. A 45 x 60 cm polythene bag is taken for the experiment.
2. The green crop chosen for ensiling is chaffed to 2-4 cm in length and wilted to 35% Dry matter.
3. The crop is thoroughly mixed. Molasses can be added to ensure the availability of soluble carbohydrate for efficient bacterial fermentation to an extent of 3 to 6.5%. Salt and urea are added at 0.5 and 1%, respectively, to improve palatability and nitrogen content of silage.
4. The material is then tightly packed into the bags.
5. The open end is sealed perfectly and kept in an identical bag and placed in a metal can/cement tub after sealing.
6. The remaining space is filled with sand before closing the lid, thus minimizing the possibility of air, permeating the polythene film.
7. The silo is then stored in a room and opened for sampling for 4-6 weeks.

15.2 Silage pit/silo preparation

A silo is an airtight structure designed for the storage and preservation of high moisture feed as silage. It may be like pit, trench, tower or stack. Pit silos are more common in India. The pits are dug 2.4 to 3.0 meter deep, with variable sizes. One cubic meter of space is required for 400 kg fodder.

The container in which silage is made is of greatest importance and determines to a large extent the nature and quality of the final product. The size of the silo generally depends upon the number and kind of animals to be fed on it and its height on the length of feeding period.

Requisites of a silo

1. The walls should be impermeable so the water can't gain entry into the silo pit. Walls may be made of cement or brick and mortar.
2. Silo should be sufficiently deep. It should not be shallow. The depth depends on the water table in the locality.
3. Silo must be located on an elevated ground.
4. The size of the silo should be calculated on the basis of the number of animals to be fed, the length of the feeding period.
5. It is useful to divide large concrete silos into smaller compartments.
6. The wall of silo must be smooth to not getting air trapped inside.
7. The walls should be strong and rigid to withstand the outward pressure exerted by the silage when it settles.

Part II

Applied Animal Nutrition (Ruminants, ANN-121)

Exercise No. 16

Conduction of Digestion Trial in Ruminants

Proximate analysis of the feeds and fodders does not give any idea about their nutritive value, because availability of nutrients varies according to the feed stuffs. The major loss of nutrients take place in body after ingestion of feed is through feces. Hence, it is very essential to know the actual amount of feed digested by the animals. Losses which take place after digestion are more or less comparable in different feedstuffs.

Digestibility

Digestibility of a feed can be defined as that proportion of feed which is not being excreted in the faeces and hence assumed to be absorbed by the animal. It is expressed as percentage (% digestibility or digestibility coefficient).

$$\% \text{ Digestibility} = \frac{\text{Absorbed (Intake - Outgo)}}{\text{Intake}} \times 100$$

Digestibility of dry matter as well as each nutrient may be determined.

For Example, if a cow ate 12 kg of hay (85% DM) supplying 10.2 kg of DM and excreted 4.2 kg DM in its faeces, the digestibility of hay DM would be 10.2-4.2/10.2 X 100 = 58.82%.

The digestibility may be of two types. Apparent digestibility and true digestibility. When digestibility is determined by subtracting the total amount of feces or nutrient present in the feces from the amount of feed or nutrient intake, then it is called apparent digestibility. Because in this case, faecal portion also include the metabolic portion (sloughed mucosa, unspent enzymes, undigested microorganisms) of body origin, besides undigested portion of feed. When metabolic portion present in the feces is subtracted from it, then the digestibility obtained is called as true digestibility and hence the true digestibility is always higher to that of apparent digestibility.

$$\text{Apparent digestibility (\%)} = \frac{\text{Intake (kg) - Total outgo in faeces (kg)}}{\text{Intake (kg)}} \times 100$$

$$\text{True digestibility (\%)} = \frac{\text{Intake (kg) - (Total outgo in faeces (kg) - faecal losses from endogenous origin (kg)]}}{\text{Intake (kg)}} \times 100$$

In general the digestibility coefficients are estimated only for organic matter. For ash or minerals it is not determined because firstly the inorganic nutrient does not contribute to energy content of the feed and secondly most of the absorbed minerals are excreted normally through the gut as endogenous portion and determination of that fraction is difficult.

Digestibility measurement through digestion trial

Selection of animals

As much as possible animals should be of same breed, age, sex (usually male are preferred over females, due to less chances of fecal samples get contamination by urine) and body weight. A minimum of four adult animals are needed for testing a diet. However, more will be the number of animals more will be the accuracy of result.

Animal should be healthy and free from parasitism, hence before starting the trial, animals should be administered anti-parasitic drugs.

Total experimental period may be divided into followings:

1. Adaptation period/ Preliminary period

This period is given to animals just to acclimatize the animal and its gastrointestinal tract for a specific type of diet or feed being tested and also to flush out the residues of the previous feed. This period mainly depends upon the species of animal being used for digestion trial. In ruminants minimum duration for this period is of three weeks (21 days).

2. Collection period

During collection period, time of offering feed and collection of sample should be strictly followed at same time daily, for full length of the collection period. In this period, amount of feed offered and left over feed (taken next day) is accurately measured. Similarly the feacal samples are quantitatively collected on 24 hours basis. Collection period in ruminants is extended from 7-10 days. During this period the representative homogeneous samples of feed offered, feed left over and feces voided are preserved and analysed for dry matter and nutrient composition.

Direct determination of digestibility of individual feed may be done if itself it is maintenance or production type e.g. Non-leguminous greens as well as their hay acts as maintenance type while leguminous greens and hays act as productive type. While for some of the feeds which are non-maintenance type (straws, stovers, dried grasses etc.) or which lack bulk and desired quantity of fiber (cereals, oil seed cakes etc.) can not be fed alone and their digestibility cannot be determined directly and requires indirect method, where digestibility is determined by difference method. Determination of digestibility by difference method may not be very accurate since addition of oil seed cakes to basal diet may influence its digestibility because of the associative effect of the feed. In addition of these two methods another method is indicator method where quantitative measurement of feed intake and feaces voided is not required for determining the digestibility and here determination of concentration of internal (Lignin, acid insoluble ash or silica) or external indicator (chromic oxide) in feed as well as feaces will be sufficient to determine the digestibility. In this method assumption is made that the indicator is excreted uniformly and therefore even small feces collected at any time during digestion trial would be sufficient to indicate the amount of nutrient present in the feces per unit of indicator.

Exercise: A cow was fed on an average 10 kg of berseem hay daily in a digestion trial and excreted 20 kg of dung. The chemical composition (% on DM basis) of the berseem hay and faeces is given in the table below, so find out the digestibility of dry matter, crude protein, ether extract, crude fiber and nitrogen free extract.

	Dry matter	Ash	CP	CF	NFE	EE
Hay	85	5	20	25	47	3
Faeces	20	25	10	40	24	1

Solution : Firstly to find out the digestibility of dry matter, we have to calculate the intake and outgo of the dry matter:

Dry matter of the berseem is 85% and intake on as such basis is 10 kg, so

Out of 100 parts 85 parts is dry matter

So in 1 part   85/100

So in 10 kg   (85/100) x 10 = 8.5 kg will be the dry matter intake

Similarly feaces out go is 20 kg and dry matter is 20% so

Out of 100 parts 20 parts is dry matter

So in 1 part   20/100

So in 20 kg   (20/100) x 20 = 4 kg is the dry matter outgo

Now the amount of DM digested will be determined as DM intake - DM outgo (8.5 kg - 4 kg) = 4.5 kg

Thus, the dry matter digestibility will be = DM digested/ DM intake x 100 i.e. (4.5/8.5) x 100 = 52.94%

Similarly the digestibility of other nutrients will be determined by obtaining the digested amount of specific nutrient and its intake and as the values of other proximate contents is given on dry matter basis, so that will be calculated on the basis of dry matter intake and outgo in following manner:

The dry matter intake is 8.5 kg and in that dry matter CP is 20%, so

Out of 100 parts 20 parts is protein

So in 1 part 20/100

So in 8.5 kg (20/100) x 8.5 = 1.7 kg is CP intake

Similarly dry matter out go in feaces is 4 kg and CP in which is 10%, so

Out of 100 parts 10 parts is dry matter

So in 1 part 10/100

So in 4 kg (10/100) x 4 = 0.4 kg is CP outgo

Now the amount of CP digested will be determined as CP intake - CP outgo (1.7 kg - 0.4 kg) = 1.3 kg

Thus, the CP digestibility will be = CP digested/ CP intake x 100 i.e. (1.3/1.7) x 100 = 76.47%

Similarly the digestibility coefficients for other nutrients will be:

Daily average	DM	CP	CF	NFE	EE
1. Intake (g)	8500	1700	2125	3995	255
2. Outgo (g)	4000	400	1600	960	40
3. Digested (g) (1-2)	4500	1300	525	3035	215
4. Digestibility Coefficient (%)	52.94	76.47	24.70	75.97	84.31

Answer : The digestibility coefficient of DM is 52.94%, for CP 76.47%, for CF 24.70%, for NFE 75.97% and for EE it is 84.31%.

Exercise No. 17

Conduction of Metabolism Trial to Determine Nitrogen Retention

Digestion trial requires feaces collection only but when we have to go for balance studies we have to collect urine also (in addition to the nutrient losses take place through milk in lactating animals) to find out the nutrient losses takes place through urine. As we have discussed in digestion trial also, male animals are usually preferred to perform these types of studies to obtain contamination free urine and feaces to find out clear picture.

Balance studies are usually performed to find out the balance of nitrogen (Nitrogen retention, NR) and minerals (Ca, P etc.) as urine is rich in these nutrients and hence for their balance we have to find out the losses takes place through urine.

Metabolism stall or cages

Balance studies mainly require the cages known as metabolism stall or cages. These are the cages usually made up of iron and sometimes wooden (especially in case of trace element balance studies) also and designed specifically to collect urine and feaces separately. The cages are so narrow that animal can only stand and sit down easily but can't turn inside them. In front side, the cages are supplied with trough for offering measured quantity of feeds and also at its one side the facility for providing water is their, from where water is offered to animals at definite intervals.

In floor the cages are provided with central grills to pass out the urine directly through gravity into the trey, which is situated beneath the grills and from a hole in the trey the urine is directed automatically in the bottles or buckets placed inside the hole. The feaces is regularly collected manually from back side of floor of cages just after defecation, by an attendant.

By obtaining the intake of nitrogen and outgo through feaces and urine the amount of nitrogen retained in the body may be calculated. The nitrogen retention is also known as nitrogen balance and which may be either positive, negative or even zero balance. When animal is given balanced diet in growing stage there will be positive nitrogen balance. Negative nitrogen balance may be obtained in the condition when poor quality of feed is available to animal, which is not sufficient to supply nitrogen as per the requirements and in that case body reserve is utilized for maintenance of body, provides negative nitrogen balance. Similar condition may also be observed in diseases when the intake of animal is reduced but outgo remained maintained. Zero nitrogen balance may be obtained in the case when adult animal is maintained at maintenance ration where only required amount of the nitrogen is fed to animals and no nitrogen gets deposited in the body because the nitrogen intake and outgo remained equal.

Nitrogen balance per day in grams may be calculated from following formula:

N balance (g/day) = Intake of N (g) - (N outgo through feaces + N outgo through urine)

Exercise: Find out the nitrogen retention (NR) from the data of exercise no 1 when the crude protein excreted in urine per day is 1000g and the cow is not producing any milk.

Solution: With the help of previous exercise it is clear that the intake of crude protein is 1700 g per day and outgo through feaces is 400 g and as mentioned in this exercise the outgo through urine is 1000 g per day.

Thus the total outgo (urinary and feacal) from body is 400 + 1000 = 1400 g per day that means out of 1700 g CP intake outgo is 1400 g per day, So the retention in body is 300 g CP per day and when we convert this data in to nitrogen then retention of nitrogen will be CP/ 6.25 (300/6.25) i. e. 48 g.

Answer: The nitrogen retention (NR) will be 48 g per day.

Exercise No. 18

Calculation of Nutritive Value of Feedstuffs (In Terms of DCP, TDN and SE)

The nutritive value of feedstuffs is generally expressed in terms of protein and energy value. The protein and energy value usually referred in ruminants are DCP (Digestible crude protein) and TDN (Total digestible nutrients), because the feeds used in the diet of ruminants have a vide variability in their digestibility, hence after removal of the variable fraction the requirement of nutrients is expressed. Starch Equivalent (SE) is also the measure of energy but its usefulness is limited to certain extent, due to involvement of tedious and expensive process.

Digestible crude protein (DCP)

The digestible crude protein (DCP) of a feedstuff expresses the amount of crude protein which gets digested after ingestion of 100 g of that specific feedstuff or the diet. To find out the DCP content of a feed, multiply the digestibility coefficient of the crude protein with the crude protein content of the feedstuff.

$$\text{DCP (\%)} = \frac{\text{CP\% of the feed x digestibility co-efficient of CP}}{100}$$

Total digestible nutrients (TDN)

As compared to DCP the word TDN as such is not correct, because as previously told that TDN is the measure of energy and hence multiplication by a factor 2.25 takes place in the case of digestible ether extract values, as the energy content of the fat is 2.25 times higher to that of carbohydrates. But in other sense it is the total of all other digestible nutrients. To determine the TDN content of a feed, the percentage of each nutrient of that feed is multiplied by its respective digestibility coefficient and added together.

TDN (%) = % DCP + % DCF + % DNFE + (% DEE x 2.25)

TDN value of feedstuffs can be used to derive other energy values, by using the following relationship:

1 kg TDN = 4.4 Mcal DE or 3.6 Mcal ME

Exercise: With the help of data of exercise no 1 find out the DCP and TDN values of the berseem hay fed to the cow.

Solution: The digestible crude protein (DCP) is the protein digested from 100 g of feed; here the feed is berseem hay.

From the exercise number 1 we find out that the digestibility of protein is 76.47% i. e. out of 100 g protein 76.47 g protein gets digested. But when we have to find, out of 100 g of feed (berseem hay) how much protein is digested so that will be DCP.

So out of 100 g protein 76.47 g protein gets digested

Out of 1g .... 76.47 / 100

So out of 20 g protein (protein present in 100 g of berseem hay) = (76.47 /100) x 20 = 15.29

So DCP% of the berseem hay is 15.29%

Now Calculation of TDN, as previously said that it is the measure of energy that's why EE is multiplied by factor 2.25. So to obtain the total digestible nutrients we have to sum up the different digestible nutrients:

DCP = 15.29, similarly the different digestible nutrients will be calculated in the following manner:

DCF = (24.70 /100) x 25 = 6.18

DNFE = (75.97 /100) x 47 = 35.71

DEE = (84.31 /100) x 3 = 2.53 (x2.25) = 5.69

So TDN % = % DCP (15.29) + % DCF (6.18) + % DNFE (35.71) + % DEE x 2.25 (5.69) = 62.87

So TDN % of the berseem hay is 62.87%

Starch equivalent (SE)

The concept of starch equivalent was proposed by Oskar Kellner. It is the fat producing power of a feed. It is the number of kilograms of starch that would be required to produce the same amount of fat as produced by 100 kg of the feed over and above the maintenance ration. For example, if we say SE of linseed cake is 74, it means that 100 kg of linseed cake can produce the same amount of fat as produced by 74 kg of pure starch when fed in addition to maintenance ration. It is actually the measure of net energy of the feedstuff. However, net energy is expressed in terms of calories and SE is expressed in terms of starch values.

For the calculation of SE, fat producing power of each of the digestible nutrient is calculated by using appropriate factors and upon addition, the total fat producing power i.e. SE of the feedstuff is known. The fat producing power of various nutrients is given below:

Nutrient	SE factor
Digestible Protein	0.95
Digestible crude fiber	1.0
Digestible NFE	1.0

Ether extract fractions from different feedstuffs differ in the fat producing power, thereby giving variable SE factors as below: -

Digestible Ether extract from roughages have factor 1.91 from cereals or their byproducts it is 2.12 and from oil cakes, oil seeds or animal products it is 2.41.

However, SE value calculated as above is subject to further correction based on the feedstuffs.

1. In case of concentrates the calculated SE value obtained is multiplied by a value number or golden number. Such value numbers range from 95 to 100 in different concentrates.

   In concentrates SE value is determined as:

   SE value = Calculated SE x Value number / 100

2. In roughages correction factor is used as per the fiber content of the roughages, which will account for its expenditure of energy in chewing, mastication, digestion etc. The correction factor is multiplied with the crude fiber content of the feed and the value obtained is deducted from the calculated SE value. The correction

factors for dry and green roughages are given in the table on next page.

The SE value for dry and green roughages obtained as:

SE value = Calculated SE value – (% CF in feed on fresh basis x respective factor)

Table : Correction factors for dry and green roughages

Feeding stuffs	Fibre % on wet basis	Crude fibre % to be multiplied by the factor
Dry roughages, long (straws, hays etc.)		0.58
Dry roughages (finely chopped)		0.29
Green fodders	4	0.29
	5	0.31
	6	0.34
	7	0.36
	8	0.38
	9	0.40
	10	0.43
	11	0.45
	12	0.48
	13	0.50
	14	0.53
	15	0.55
	16 or more	0.58

Exercise : Find out the starch equivalent (SE) of the Barley grains (value no 98), having 7.6 % DCP, 1.2 % DEE, 60.9 % DNFE and 2.5 % DCF.

Solution : Starch equivalent (SE) may be calculated by digestible nutrients after multiplying with their respective factors in the following manner:

Nutrients	% Digestible nutrients	Factor	SE value
CP	7.6 (DCP)	0.95	7.22
EE	1.2 (DEE)	2.12	2.54
NFE	60.9 (DNFE)	1	60.9
CF	2.5 (DCF)	1	2.5
Calculated SE			**73.16**

Now calculated SE value will be multiplied with the value no % i. e. 73.16 x 98 /100

SE = 71.70

Answer : The starch equivalane (SE) value of barley will be 71.70

Exercise: Find out the starch equivalent (SE) of the green berseem, having 2% DCP, 0.5% DEE and 9.0% of digestible total carbohydrate (DCHO) and the crude fiber of the berseem is 6.0%.

Solution : Starch equivalent (SE) may be calculated by digestible nutrients after multiplying with their respective factors in the following manner:

Nutrients	% Digestible nutrients	Factor	SE value
CP	2 (DCP)	0.95	1.90
EE	0.5 (DEE)	1.91	0.96
TCHO	9.0 (DTCHO)	1	9.0
Calculated SE			**11.86**

Now correction factor for green berseem will be calculated i .e 6 x 0.34 = 2.04

Here correction factor will be deducted from calculated SE value i. e. 11.86 - 2.04

SE = 9.82

Answer : The starch equivalane (SE) value of green berseem will be 9.82

Exercise: Find out the starch equivalent (SE) of the paddy straw (crude fiber 35%), having DCP =0, DEE= 0.80, DNFE = 17.80 and DCF = 24.50%.

Solution: Starch equivalent (SE) may be calculated by digestible nutrients after multiplying with their respective factors in the following manner:

Nutrients	% Digestible nutrients	Factor	SE value
CP	0 (DCP)	0.95	0
EE	0.80 (DEE)	1.91	1.53
NFE	17.80 (DNFE)	1	17.80
CF	24.50 (DCF)	1	24.50
Calculated SE			**43.83**

Now correction factor for paddy straw will be calculated *i.e.* 35 x 0.58 = 20.30

Here correction factor will be deducted from calculated SE value *i.e.* 43.83 - 20.30

SE = 23.53

Answer : The starch equivalent (SE) value of paddy straw will be 23.53

Exercise: In the above mentioned exercise find out the SE value of chopped paddy straw.

Solution: In chopped paddy straw the correction factor will be changed i. e. 0.29 will be used in place of 0.58, while other calculations will be the same, so correction factor for chopped paddy straw will be 35 x 0.29 = 10.15

So SE will be, calculated SE (43.83) - correction factor (10.15) = 33.68

Answer : The starch equivalent (SE) value of chopped paddy straw will be 33.68

Exercise No. 19

Nutritive Value of Common Feedstuffs in Ruminant's Diet

It is very essential to know the nutritive value of common feedstuffs used in the diet of ruminants. On the basis of these values the nutrient requirements of ruminants may be fulfilled for different body functions like maintenance, growth, lactation, wool production or others. The nutritive value of various common feedstuffs used in the diet of ruminants is being presented in the table below:

Table : Nutritive value of common feed stuffs:

Feed stuffs	DM	DCP	TDN
Cereals Straws			
Rice straw	90	0	35
Wheat straw	90	0	42
Oat straw	90	0	45
Ragi straw	90	0	40
Jowar karbi	90	1.0	50
Bajra karbi	90	0.8	48
Legumes Straw			
Gram straw	90	2.0	21
Hays			
Berseem hay	90	9	60

Lucerne hay	90	14.6	50
Cowpea hay	90	9	45
Non-leguminous green			
Maize	25	1.0	16
Oat	25	2.5	16.5
Jowar	30	0.8	16
Bajra	30	1.0	14.6
Guar	20	1.3	10
Napier grass	25	0.9	13
Leguminous green			
Cowpea	20	2.6	12
Lucerne	20	3	12
Berseem	20	2.8	13
Cereal grains			
Maize	90	8	85
Oat	90	7	78
Barley	90	8	78
Jowar	90	7-8	80
Bajra	90	8	75-80
Mill by-products			
De-oiled rice bran	90	12	70-72
Wheat bran	90	10	67
Rice bran	90	8	68
Rice polish	90	12	75-76
Dal chuni	90	6-8	60
Oil cakes			
Gingilly cake/ Til cake / Sesame cake	90	38	78
Groundnut cake	90	42	71
Coconut cake	90	19	72
Cottonseed cake	90	19	72
Mustard cake	90	28	74
Linseed cake	90	28	65
Soybean meal	90	38	75
Animal protein supplements			
Fish meal	90	60	70

Exercise No. 20

Preparation of Concentrate Mixture or Calf Starter or Creep Ration

When we are having two or more than two feed ingredients and by using them we have to prepare a mixture of desired nutrient composition then we will require to use Pearson's square method. Use of pearson's square is a simple procedure that allows us to mix two feed stuffs with different nutrients concentrations and come up with a mixture of the desired total concentration. For the pearson's square to work, the desired diet' nutrient concentration must be between the nutrient concentrations of the two feed stuffs. Suppose we have a protein source such as ground nut cake (GNC) with 40.9% crude protein (CP) and maize with 8.8% CP, and we need a mixture of the two that will have 18% CP. AS shown in the illustration, we compare the CP percentage of each feed on the left with the desired percentage in the middle of the square. The lesser value is subtracted from the greater value and the answer, in parts of a mixture rather then in percentage, is recorded diagonally, but read horizontally.

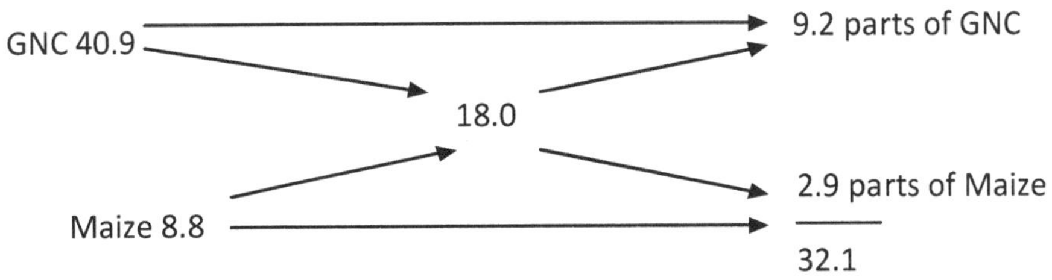

Answer

9.2 parts of GNC and 22.9 parts of maize out of 32.1 parts

% of GNC in mix = (9.2 / 32.1) x 100 = 28.66

% of maize in mix = (22.9 / 32.1) x 100 = 71.34

Check for CP

28.66% GNC x 40.9 CP/100 = 11.72%

71.34% maize x 8.8% CP/100 = 6.28%

18.00% total CP in the mix

The same procedure can be used for energy, minerals and so on. Furthermore, calories, ppm or other units of measurements can be used in place of percentage.

Sometimes we might want to have the exact amounts of the two major nutrients, such as CP and TDN for which we must have a minimum of three feedstuffs. We can accomplish this by going through three pearson's squares.

For proper functioning of pearson's square we must aware of some facts, for obtaining a desired centre value the two ingredients or mixtures which are being taken one of them should be having value higher and another lower to that of central value. It will not be possible to find out a desired value if both the ingredients having their values either higher to the central value or lower to the central value and secondly if we have to add mineral and salt with 100% concentrate mixture we have to adjust the desired figures according to the percentage of mineral and salt required to be added.

Suppose we want a final mix with 20% CP and 70% TDN. We have maize with 10% CP and 85% TDN, ground nut cake (GNC) with 40% CP and 70% TDN, wheat bran with 14% CP and 68% TDN and fish meal with 60% CP and 70% TDN. So first, we will go through two perason's squares and get two mixtures exact for one nutrient, in this example we will do CP first. Out of these two mixtures one mixture must be having with 20% CP and greater than 70% TDN, while another with 20% CP and less than 70% TDN.

1. First we will have to adjust the CP and TDN values as per the addition of mineral mixture and salt in the final concentrate mixture.

Suppose we have to add 3% mineral supplements and salt, so there will be only 97 parts of ingredients will be there, to contribute in the protein and energy requirements, so that we have to adjust the desired figure by 97%.

First for protein:

For 97 parts we have to prepare 20 % CP

Then for 1 part 20/97

So for 100 parts (20/97) x 100 = **20.62% CP**

Similarly for TDN

For 97 parts we have to prepare 70% TDN

Then for 1 part 70/97

So for 100 parts (70/97) x 100 = **72.16 % TDN**

2. Now we have to choose out of four ingredients only two ingredients at a time for adjustment of protein %

Suppose we are taking GNC and Wheat bran in first pearson for preparing mixture A for adjusting the CP:

Mix A

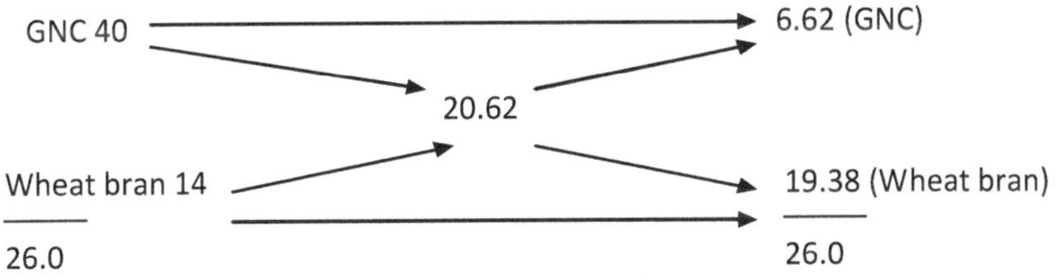

So out of 26 parts 6.62 parts should be from GNC

Then for 1 part 6.62/26

So for 100 parts (6.62/26) X 100 = 25.46% GNC

Similarly out of 26 parts 19.38 parts should be from wheat bran

Then for 1 part 19.38/26

So for 100 parts (19.38/26) X 100 = 74.54% Wheat bran

Now when we calculate the CP and TDN supplied from this composition so

	CP	TDN
25.46 Parts GNC	(25.46 x 40) / 100 = **10.18**	(25.46 x 70) / 100 = **17.82**
74.54 Parts Wheat bran	(74.54 x 14) / 100 = **10.44**	(74.54 x 68) / 100 = **50.69**
Mix A	**20.62**	**68.51**

3. Now we have to choose out of four ingredients the another two ingredients for adjustment of protein %

   Now we are taking fish meal and maize in second pearson for preparing mixture B for adjusting the CP:

   Mix B

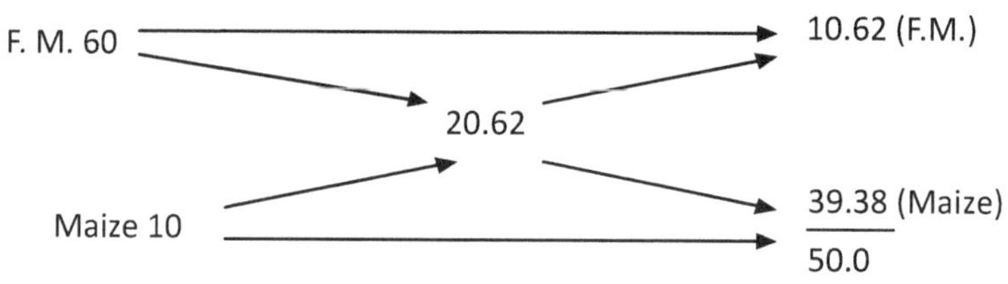

So out of 50 parts 10.62 parts should be from fish meal

Then for 1 part 10.62/50

So for 100 parts (10.62/50) X 100 = **21.24 % Fish meal**

Similarly out of 50 parts 39.38 parts should be from maize

Then for 1 part 39.38/50

So for 100 parts (39.38/50) X 100 = **78.76% maize**

Now when we calculate the CP and TDN supplied from this composition so

	CP	TDN
21.24 Parts FM	(21.24 x 60) / 100 = **12.74**	(21.24 x 70) / 100 = **14.87**
78.76 Parts Maize	(78.76 x 10) / 100 = **7.88**	(78.76 x 85) / 100 = **66.95**
Mix A	**20.62**	**81.82**

4. We have adjusted CP by making two mixtures A and B, now we have to adjust TDN with the help of these two mixtures having TDN % of 68.51 and 81.82 %, respectively.

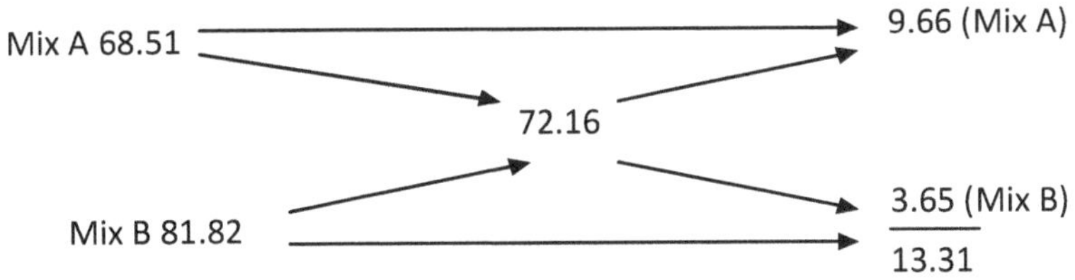

So out of 13.31 parts 9.66 parts should be from mixture A

Then for 1 part 9.66/13.31

So for 97 parts (3% Mineral and salt) (9.66/13.31) X 97 = **70.40% Mixture A**

Similarly out of 13.31 parts 3.65 parts should be from mixture B

Then for 1 part 3.65/13.31

So for 97 parts (3.65/13.31) X 97 = **26.60% mixture B**

5. Now when we calculate the required amount of individual ingredients in the concentrate mixture

   GNC

   Out of 100 parts GNC is 25.46 parts

   Then for 1 part 25.46/100

   So for 70.40 parts (25.46/100) X 70.40 = **17.92% GNC**

   Wheat bran

   Out of 100 parts wheat bran is 74.54 parts

   Then for 1 part 74.54/100

   So for 70.40 parts (74.54/100) X 70.40 = **52.48% Wheat bran**

   Fish meal

   Out of 100 parts fish meal is 21.24 parts

   Then for 1 part 21.24/100

   So for 26.60 parts (21.24/100) X 26.60 = **5.65% Fish meal**

   Maize

   Out of 100 parts maize is 78.76 parts

   Then for 1 part 78.76/100

   So for 26.60 parts (78.76/100) X 26.60 = **20.95% Maize**

6. Now we have to check out the amount of CP and TDN supplied by complete concentrate mixture:

Ingredients Amount (kg)	Supplied CP	Supplied TDN
GNC 17.92	(17.92 x 40)/100 = **7.17**	(17.92 x 70)/100 =**12.54**
Wheat bran 52.48	(52.48 x 14)/100 = **7.35**	(52.48 x 68)/100 =**35.69**
Fish meal 5.65	(5.65 x 60)/100 = **3.39**	(5.65 x 70)/100 = **3.96**
Maize 20.95	(20.95 x 10)/100 = **2.10**	(20.95 x 85)/100= **17.81**
Mineral and salt 3.00	-	-
Total 100.00	20.01	70.00

So it is clear that this procedure works very well to produce a mix with exact specifications for two nutrients. If we want to add more nutrients, it will require more than three squares. But diets that have exact requirements for more than two nutrients are tedious to formulate by hand.

Exercise No. 21

Nutrient Requirements for Dairy Cattle and Buffaloes

Nutrient requirements of the animals in terms of protein, energy, minerals and vitamins are the basis for scientific feeding. The total requirement of an animal includes two components (i) maintenance (ii) production. While maintenance requirement varies with body size of the animal, the requirement for production depends upon its physiological state and production levels.

Such recommendations are based on the results of a large number of feeding experiments, conducted throughout the world. Feeding standards developed by Nation Research Council (NRC) of USA and Agricultural Research Council (ARC) of UK, for livestock and poultry are considered to be the best general guides for computation of ration. However, limited investigations carried out in this country suggest that adoption of foreign standards always leaves a fair margin of safety so far as the feeding of Indian species are concerned. This owes mainly due to differences in size, breed, nutritional values of feeding stuffs and other practices in the agro-climatic conditions of this tropical country. Some efforts have been made by India scientists in this area.

The feeding standard given by ICAR (1998) in terms of DM, DCP, TDN, ME, Ca, P and carotene intake of animals are based on the recommendations of scientific panel on Animal Nutrition and

Physiology. This standard includes recommendation based on experimental work carried out in India over the past several years.

Nutrient requirements for maintenance:

For this mature and dry animals of same sex and body weight were fasted, kept in thermo-neutral environment and their fasting heat production was determined. The fasting metabolism gives an estimate of the minimum quantity of net energy required for maintenance. The studies conducted in India are mainly based on indirect calorimetry. On the basis of several studies conducted on dry and lactating animals the energy requirement for maintenance is calculated using 122 kcal of ME / $W^{0.75}$ for cattle and buffalo. Similarly, protein requirements were generated by using factorial method, where endogenous urinary nitrogen (EUN) and metabolic fecal nitrogen (MFN) were estimated and on that basis 2.84 g of DCP/ $W^{0.75}$ requirement was established. Calculated values on that basis are represented in the following table.

Table: Maintenance requirements of the mature cattle and buffaloes (ICAR, 1998)

Body weight (Kg)	Dry feed (Kg)	DCP (g)	TDN (Kg)	ME (Mcal)	Ca (g)	P (g)	Carotene (mg)
200	3.5	150	1.7	6.0	8	7	21
300	4.5	200	2.4	8.4	12	10	32
400	5.5	250	3.0	10.8	17	13	42
500	6.5	300	3.7	13.2	20	15	53
600	7.5	350	4.2	15.5	22	17	64
700	8.5	390	4.8	17.3	25	19	74
800	9.5	430	5.3	19.1	27	21	85

Nutrient requirements for pregnancy

No systemic work for requirements of pregnant cattle and buffalo has been conducted in India and hence the recommendations of NRC (1989) are being followed and the table of the same is presented below. These recommendations are to be followed in third trimester of pregnancy, while before that no additional nutrient requirements are considered.

Table: Nutrient requirements of the cattle and buffaloes during last trimester of pregnancy:

Body weight (Kg)	Dry feed (Kg)	DCP (g)	TDN (Kg)	ME (Mcal)	Ca (g)	P (g)	Carotene (mg)
300	5.6	290	3.4	12.4	16	14	56
400	7.2	350	4.0	14.1	23	18	76
500	8.6	430	4.8	17.3	29	22	95
600	10.0	500	5.6	20.2	34	26	114
700	11.3	550	6.3	22.7	39	30	133
800	12.6	630	7.1	25.6	44	34	152

Nutrient requirements for lactation

In practical feeding many factors like level of intake, proportion of concentrate to roughage and the quality of roughage, physical form of the ration (grinding, pelleting etc.) will affect the energy requirement. The requirements for TDN and DCP have been calculated per kg of milk production taking 1188 kcal of ME per kg of 4% fat corrected milk (FCM) and 132 g digestible nitrogen for 100 g of milk nitrogen.

Table: Nutrient requirements per kg of milk production (ICAR, 1998):

Fat (%)	DCP (g)	TDN (g)	ME (Mcal)	Ca (g)	P (g)
3	40	270	0.97	2.5	1.8
4	45	315	1.13	2.7	2.0
5	51	370	1.28	2.9	2.2
6	57	410	1.36	3.1	2.4
7	63	460	1.54	3.3	2.6
8	69	510	1.80	3.5	2.8
9	75	550	2.06	3.7	3.0
10	81	600	2.16	3.9	3.2
11	85	650	2.34	4.1	3.4

Nutrient requirements for growth

Nutrient requirements for growing calves may be divided into two phases. First is the period of pre-ruminant growth period where the anatomical and physiological development of rumen does not takes place completely and thus the microbial fermentation did not play a major role in the nutrient requirements of the animals and hence protein quality plays a major role in the growth of these calves.

Table: Nutrient requirement of pre ruminant cattle and buffalo calves (ICAR, 1998)

Age (days)	B. wt. (kg)	Daily Gain (g)	DCP (g)	TDN (kg)	ME (Mcal)	Ca (g)	P (g)	Vit-A 100IU
0-15	25	200	80	0.40	1.5	2.5	1.5	1.5
16-30	30	300	90	0.50	1.7	3.0	2.0	1.5
31-60	40	300	125	0.80	2.4	3.5	2.5	1.7
61-90	50	350	150	1.00	3.6	4.0	3.8	2.0

In the second phase, when the development of rumen takes place completely and quality of nutrient gets modified by the help of microbial fermentation, hence the protein quality did not remains a major concern as microbial protein works as a major portion of protein supply through diet.

The requirement for growing cattle/buffalo calves may be considered by using following tables as per the expected growth rate required to be obtained:

Table : Nutrient requirement for growing calves (cattle, ICAR, 1998)

B Wt. (kg)	Growth rate (g)	DM (Kg)	DCP (g)	TDN (Kg)	ME (Kcal)	Ca (g)	P (g)	Carotene (mg)	Vit-A (100 IU)
70	550	2.1	255	1.4	5.0	8	6	7	3
	600	2.3	260	1.5	5.4	8	6	7	3
80	550	2.4	260	1.5	5.4	9	7	7	3
	600	2.5	270	1.6	5.8	9	7	8	3
100	550	2.8	270	2.1	7.6	10	8	10	3
	600	3.0	280	2.2	7.9	10	8	10	3
120	550	3.2	280	2.3	8.3	14	11	11	4
	600	3.7	290	2.4	8.6	14	11	11	4
140	550	3.7	290	2.5	9.0	14	11	16	6
	600	4.2	300	2.6	9.4	14	11	16	6
160	550	4.2	300	2.7	9.7	15	12	16	6
	600	4.7	320	2.8	10.1	15	12	16	6
180	550	4.7	310	2.9	10.4	15	12	18	8
	600	5.2	320	3.0	10.8	15	12	18	8
200	550	5.2	320	3.1	11.2	16	13	22	9
	600	5.7	330	3.2	11.5	16	13	22	9
220	550	5.7	330	3.3	11.9	18	14	24	10
	600	6.2	340	3.4	12.2	18	14	24	10
240	550	6.2	340	3.5	12.6	20	16	28	12
	600	6.7	350	3.6	13.0	20	16	28	12
260	550	6.7	350	3.7	13.3	20	16	30	13
	600	7.2	360	3.8	13.7	20	16	30	13

Table: Nutrient requirements for growing calves (buffaloes, ICAR, 1998)

B Wt. (kg)	Growth rate (g)	DM (Kg)	DCP (g)	TDN (Kg)	ME (Kcal)	Ca (g)	P (g)	Carotene (mg)	Vit-A (100 IU)
70	450	2.0	156	1.3	4.7	8	6	7	3
	550	2.1	168	1.5	5.4	8	6	7	3
80	450	2.2	177	1.4	5.0	9	7	7	3
	550	2.3	183	1.6	5.7	9	7	8	3
100	450	2.7	198	1.7	6.1	10	8	10	3
	550	2.8	210	1.8	6.5	10	8	10	3
120	450	3.1	279	1.9	6.8	14	11	11	4
	550	3.2	291	2.1	7.6	14	11	11	4
140	450	3.6	300	2.2	7.9	14	11	16	6
	550	3.7	316	2.2	7.9	14	11	16	6
160	450	4.1	264	2.5	9.0	15	12	16	6
	550	4.2	276	2.6	9.4	15	12	16	6
180	450	4.6	288	2.7	9.9	15	12	18	8
	550	4.7	300	2.8	10.1	15	12	18	8
200	450	5.1	312	2.9	10.4	16	13	22	9
	550	5.2	329	3.1	11.2	16	13	22	9
220	450	5.6	345	3.1	11.2	18	14	24	10
	550	5.9	357	3.2	11.5	18	14	24	10
240	450	6.1	375	3.3	11.9	20	16	28	12
	550	6.2	390	3.4	12.2	20	16	28	12
260	450	6.2	410	3.5	12.6	20	16	30	13
	550	6.5	420	3.6	12.8	20	16	30	13

Nutrient requirements for breeding bulls

When bulls have developed properly, they are normally fed on maintenance requirement. Overfeeding should be avoided as it leads to excessive fattening, which lowers the libido. No systemic work on nutrient requirements of breeding bulls had been conducted in the country. However, the requirements proposed by Ranjhan (1980) have been adopted by ICAR (1998).

Table: Nutrient requirements for breeding bulls (ICAR, 1998)

B. wt. (kg)	DCP (g)	TDN (kg)	ME (Mcal)	Ca (g)	P (g)	Carotene (mg)
400	380	3.6	13	18	13	40
500	450	4.5	16.2	20	15	53
600	530	5.4	19.4	22	17	64

Nutrient requirements for working bullocks

Work production normally does not involve extra protein over the maintenance. The requirement of nutrients for bullocks for work production is comparatively lower than for fast moving horses.

For bullocks the work has been categories in two types:

a. *Normal work*: Working for 4 hours of ploughing or 6 hours of carting.

b. *Heavy work*: Working for 6 hours of ploughing or 8 hours of carting.

Table: Nutrient requirements for working bullocks (ICAR, 1998)

B. wt.	Normal work				Heavy work			
(kg)	DM (Kg)	DCP (g)	TDN (Kg)	ME (Mcal)	DM (Kg)	DCP (g)	TDN (Kg)	ME (Mcal)
200	4.0	240	2.0	7.2	5.0	250	2.7	9.5
300	5.8	330	3.1	11.4	7.0	420	4.0	14.4
400	7.6	450	4.0	14.4	9.8	570	4.8	17.3
500	9.4	560	4.9	18.0	11.2	710	6.4	23.0
600	11.2	660	5.8	20.8	13.4	820	8.0	28.8

Exercise No. 22

Formulation of Rations for Dairy Cattle and Buffaloes

The formulation of ration for various categories of livestock depends upon the nutrient requirements for a particular animal or group of animals and the feeds and fodders available for the purpose. Livestock feeds can be grouped into two major categories - Roughages and concentrates, on the basis of bulkiness and chemical composition and combination of both is essential for a balanced ration.

Roughages are the bulky feeds having crude fiber more than 18% and TDN less than 60% while, Concentrates are feed or feed mixture having crude fiber less than 18% and TDN more than 60%. They are rich in either energy or protein and accordingly are called as energy concentrates (when CP is less than 18%) or protein concentrates (when CP value exceeds 18%). Various energy concentrates are grains and seeds (maize, barley, sorghum etc.), milling byproducts (brans and chunis), roots (tapioca tubers, turnip, potatoes) and molasses.

Protein supplement, on the other hand may be further categorized according to the source of origin (a) animal protein supplements (meat meal, blood meal, fish meal, feather meal *etc.*) and (b) plant protein supplements (oilseed cakes and meals i.e. groundnut cake, sesame cake, mustard cake, linseed oil meal *etc.*).

The nutrient requirements of an animal may be divided in to two parts first is maintenance and another is the production requirements.

The ration, which just enables the animal at rest to carry on the essential processes of life such as breathing, circulation of blood etc. without either gain or loss of body weight is known as maintenance ration. It is dependent upon live weight of animal and also on the level of production. At higher level of production even at the same body weight, the maintenance requirements get increased. The production ration may be in the form of milk production, growth in case of growing animals, pregnancy allowances (in last trimester of pregnancy) in pregnant animals and even fat deposition in case of fattening animals.

Balanced ration

A balanced ration is one that furnishes nutrients in such proportions and amounts that will properly nourish a given animal for 24 hours. In addition, the required nutrients must be contained in the amount of DM the animal is able to consume (DMI) in the 24 hr period.

Formulation of ration is therefore, done in three steps

1. To calculate the probable dry matter intake of the animal
2. To calculate the nutrient requirements of the animal
3. To find out the amount of available feedstuffs required to fulfil the animal's nutrient requirements within its expected DMI.

1. Calculation of DMI (probable)

The dry matter intake of the animal is highly variable. It depends mainly on body weight of animal and its production (growth, fattening, milk production or pregnancy). Normal range of DM consumption in cattle and buffaloes is as follows:

Cattle	2.0 to 2.5% of live weight
Buffalo	2.5 to 3.0% of live weight

In animals with high milk production DMI is more and the level of DMI may be increased up to 3.5 to 4% and even 5% levels with high producing animals.

2. Calculation of nutrient requirements (DCP and TDN)

The nutrient requirements of the animal is calculated in terms of DCP and TDN for maintenance in case of adult unproductive animals (derived directly from table as per the body weight of the animal), while for milk production maintenance requirement (as per the body weight) will be added in the production requirements (as per the level of production, the requirements is considered for 4% milk fat in cow's milk, while 7% in case of buffalo milk) for obtaining the complete requirements of lactating animals, while the requirement for advanced stage of pregnancy, growth stage (for desired weight gain), working bullocks and breeding bulls is directly taken from the tables as per the body weight of the animals.

3. Providing the available feed stuffs to fulfil the requirements

Bulk is specifically important in the diet of ruminants. To provide the required bulk in the diet, the DM allowance for cattle and buffalo should be divided as follows:

The required dry matter will be divided into concentrate and roughages. The amount of concentrate will be $1/3^{rd}$ of the total DM supply, than remaining $2/3^{rd}$ will be supplied through roughages. The distribution of DM in dry and green roughages will depends on the availability of green fodder. If the green fodder available is leguminous then the DM supplied through that will be $1/4^{th}$ of the total DM supplied through fodder and remaining $3/4^{th}$ will be supplied through dry fodder like wheat straw. In case when green fodder available is non-leguminous then the DM supplied through that will be $1/3^{rd}$ of the total DM supplied through fodder and remaining $3/4^{th}$ will be supplied through dry fodder like wheat straw.

Computation of ration when green roughages are not available

The availability of green roughages is a difficult task throughout the year. Green fodder usually available in very limited period of time, especially in the rainy season only and sometimes also it remains available in limited quantity. In that case the supply of nutrients has to be done with the help of concentrate mixture and wheat straw only. Here wheat straw will supply only TDN, as its have zero DCP value, hence in this type of ration formulations all the DCP will have to be supplied through the concentrate mixture.

In this condition we have to find out the nutrient requirements of the animal and after that all the DCP requirements will have to be fulfilled with the help of concentrate mixture and then wheat straw will made available to the animal free choice and the animal will consume the straw as per its requirements.

Computation of ration in excess of greens

Greens are the cheapest source of nutrients and when green fodder is available in excess then it may also serves as a replacement of expensive concentrate mixture. The non leguminous greens are usually considered as maintenance type roughage and in excess they may be fed to adult dry animals free choice with little or no wheat straw and if leguminous green is available, then it may even replace the concentrate mixture for production purpose also as it referred as production roughage. Care should be taken for using green berseem as it is rich in saponins and if fed alone it may create the problem of tympani, so it should properly be used with dry fodder in combination.

Exercise : Formulate balanced ration for a cow weighing 400 kg and producing 6 litres of milk, when the available feed stuffs are green maize, wheat straw and concentrate mixture (DCP 12% and TDN 60%).

Solution **:** First of all we have to determine the probable dry matter intake of the cow, for that we have to consider @ 2.5% of live weight *i.e.* 2.5 kg for each 100 kg so for 400 kg: = 4 x 2.5 = 10 kg

Now the DM supply through concentrate mixture will be ($1/3^{rd}$) = 10 x 1/3 = 3.3 kg

Similarly as roughages (either 10 -3.3=6.7 kg or 10 x 2/3 = 6.7 kg)

Now out of the total roughage DM (6.7 kg), DM supply through green roughage (maize, non leguminous, $1/3^{rd}$ of the dry matter of the roughage) will be

= 6.7 x 1/3 = 2.2 kg

Similarly as dry roughages (either 6.7-2.2 = 4.5 kg or 6.7 x 2/3 = 4.5 kg).

Now in *second step* we have to calculate the total nutrient requirements of the animal *i.e.*

Requirements	**DCP (g)**	**TDN (kg)**
For maintenance of 400 kg body weight	250	3.0
For production of 1 lit milk (4% fat)	45	0.316
For production of 6 lits milk.	45 x 6 = 270	0.316 x 6 = 1.90
Total requirements	**520**	**4.90**

In the *third step* we have to supply the nutrients through available feedstuffs in DMI capacities:

Feedstuffs	**DM (kg)**	**As fed basis (kg)**	**DCP (g)**	**TDN (kg)**
Concentrate mixture (90, 12, 60)	3.3	3.63	436	2.18
Green maize (25, 1, 16)	2.2	8.80	88	1.41
Wheat straw (90, 0, 42)	4.5	4.95	0	2.08
Total	**10**	**17.38**	**524**	**5.67**

Answer: We have to fed concentrate mixture 3.63 kg, green maize 8.8 kg and wheat straw 5 kg to fulfill the requirements of the cow.

Exercise: Formulate balanced ration for a buffalo weighing 500 kg and producing 5 litres of milk, when the available feed stuffs are green maize, wheat straw and concentrate mixture (DCP 11% and TDN 60%).

Solution : First of all we have to determine the probable dry matter intake of the buffalo, for that we have to consider @ 2.5% of live weight i .e. 2.5 kg for each 100 kg, so for 500 kg: = 5 x 2.5 = 12.5 kg

Now the DM supply through concentrate mixture will be ($1/3^{rd}$) = 12.5 x 1/3 = 4.2 kg

Similarly as roughages (either 12.5 - 4.2 = 8.3 kg or 12.5 x 2/3 = 8.3 kg)

Now out of the total roughage DM (8.3 kg), DM supply through green roughage (maize, non leguminous, $1/3^{rd}$ of roughage dry matter) will be = 8.3 x 1/3 = 2.8 kg

Similarly as dry roughages (either 8.3 -2.8 = 5.5 kg or 8.3 x 2/3 = 5.5 kg).

Now in *second step* we have to calculate the total nutrient requirements of the animal i. e.

Requirements	**DCP (g)**	**TDN (kg)**
For maintenance	300	3.7
For production of 1 lit milk (7% fat)	63	0.460
For production of 5 lits milk.	63 x 5 = 315	0.460 x 5 = 2.30
Total requirements	**615**	**6.00**

In the *third step* we have to supply the above nutrients within the DMI capacities of the animal:

Feed stuffs	DM (kg)	As fed basis (kg)	DCP (g)	TDN (kg)
Concentrate mixture (90, 11, 60)	4.2	4.62	508	2.77
Green maize (25, 1, 16)	2.8	11.20	112	1.79
Wheat straw (90, 0, 42)	5.5	6.05	0	2.54
Total	**12.5**	**21.87**	**620**	**7.10**

Answer: We have to fed concentrate mixture 4.62 kg, green maize 11.2 kg and wheat straw 6 kg to fulfill the requirements of the buffalo.

Exercise: Formulate balanced ration for a buffalo weighing 500 kg and producing 10 litters of milk, when the available feed stuffs are green berseem, wheat straw and concentrate mixture (DCP 12% and TDN 75%).

Solution : *First* of all we have to determine the probable dry matter intake of the buffalo, for that we have to consider @ 3% of live weight i .e. 3 kg for each 100 kg, so for 500 kg: = 5 x 3 = 15 kg

Now the DM supply through concentrate mixture will be ($1/3^{rd}$) = 15 x 1/3 = 5 kg

Similarly as roughages (either 15-5 = 10 kg or 15 x 2/3 = 10 kg)

Now out of the total roughage DM (10 kg), DM supply through green roughage (berseem, leguminous, $1/4^{th}$ of the dry matter of roughage) will be = 10 x 1/4 = 2.5 kg

Similarly as dry roughages (either 10 -2.5 = 7.5 kg or 10 x 3/4 = 7.5 kg).

Now in *second step* we have to calculate the total nutrient requirements of the animal i. e.

Requirements	DCP (g)	TDN (kg)
For maintenance	350	4.2
For production of 1 lit milk (7% fat)	63	0.460
For production of 10 lits milk.	63 x 10 = 630	0.460 x 10 = 4.60
Total requirements	**980**	**8.8**

In the *third step* we have to supply the above nutrients within the DMI capacities of the animal:

Feedstuffs	DM (kg)	As fed basis (kg)	DCP (g)	TDN (kg)
Concentrate mixture (90, 12, 75)	5	5.5	660	4.13
Green berseem (20, 2.8, 13)	2.5	12.5	350	1.625
Wheat straw (90, 0, 42)	7.5	8.25	0	3.47
Total	**15**	**26.25**	**1000**	**9.225**

Answer: We have to fed concentrate mixture 5.5 kg, green berseem 12.5 kg and wheat straw 8.25 kg to fulfill the requirements of the buffalo.

Exercise: Formulate balanced ration for a buffalo weighing 500 kg and producing 5 litres of milk, when the feed stuff available is green berseem and wheat straw.

Solution: *First* of all we have to determine the probable dry matter intake of the buffalo, for that we have to consider @ 2.5% of live weight i .e. 2.5 kg for each 100 kg, so for 500 kg: = 5 x 2.5= 12.5 kg

Now in *second step* we have to find out the nutrient requirements of the animal i. e. for maintenance of buffalo weighing 500 kg and producing 5 litres of milk in the following manner:

Requirements	DCP (g)	TDN (kg)
For maintenance of 500 kg body weight	300	3.7
For production of 1 lit milk (7% fat)	63	0.460
For production of 5 lits milk.	63 x 5 = 315	0.460 x 5 = 2.30
Total requirements	**615**	**6.00**

In the *third step* we will supply the nutrients within available DMI capacities:

Here berseem is the only source to provide the DCP, as wheat straw is having zero DCP. Now we have to supply all the DCP through berseem, so the amount of berseem required will be determined by dividing the total DCP required by DCP content of the berseem i. e. 615/28 (2.8% DCP means 2.8 g in 100 g, so 28 g in 1 kg or 1000g) = 22 kg. So 22 kg of green berseem will supply DM = (20/100) x 22 = 4.4 kg. So 4.4 kg DM supplied through berseem will fulfill all the DCP requirements of the animal.

Now we have to find out the amount of TDN supplied through 22 kg berseem.

13% TDN means 130 g in 1 kg, so in 22 kg it will be 2.86 kg.

Now the amount of TDN required to be supplied through wheat straw will be 6.00 - 2.86 = 3.14 kg and which may be easily be supplied when 8 kg of the DM will be supplied by wheat straw. Thus it is indicated that the production requirements of the animal may be fulfilled by green berseem in combination with wheat straw and when we fed the berseem it shold also be kept in mind that we cant fed sole berseem diet in the animals as it is rich in saponns and will leads to tympany so combination with wheat straw or any other dry fodder is essential.

Answer: We have to fed total of 22kg (4.4 kg DM) of green berseem, which will supply all the DCP requirements of the bufffalo and then wheat straw will be offered to the animal free choice which will take care of the requirements of TDN.

Exercise: Formulate balanced ration for a buffalo weighing 600 kg and producing 10 litres of milk, when the available feed stuffs are wheat straw and concentrate mixture (DCP 14% and TDN 75%).

Solution: First of all we have to determine the probable dry matter intake of the buffalo, for that we have to consider @ 3 % of live weight i .e. 3 kg for each 100 kg, so for 600 kg: = 6 x 3 = 18 kg

So we have to supply the required nutrients with in the range of its dry matter intake capacity.

Now in *second step* we have to calculate the total nutrient requirements of the animal i. e.

Requirements	**DCP (g)**	**TDN (kg)**
For maintenance	350	4.2
For production of 1 lit milk (7% fat)	63	0.460
For production of 10 lits milk.	63 x 10 = 630	0.460 x 10 = 4.60
Total requirements	**980**	**8.8**

In the *third step* their will be supply of nutrients through available DMI capacities and in that also we have to keep one thing in mind that their will be only concentrate mixture to supply the DCP as wheat straw dont have DCP.

So we have to find out the amount of concentrate mixture to supply complete demand of DCP i. e. 980/140 = 7 kg Now the amount of DM being supplied through 7kg concentrate mixture will be 7 x 0.9 (90/100) = 6.3 kg.

Now it is clear that 6.3 kg dry matter or 7 kg of concentrate mixture will supply complete (980 g) requirement of protein. Now we have to find out the amonut of TDN being supplied by 7 kg concentratre mixture = 7 x 0.75 = 5.25 kg

The remaining amount of will be 8.8-5.25 = 3.55 kg.

Now we have to find out the amoun of wheat straw required to fulfill the requirements of remaining TDN.

So here it is clear that their will be requirements of 3.55 kg TDN, which have to be fulfilled by wheat straw. so the amount of wheat straw required to supply 3.55 kg TDN will be 3.55/0.42 = 8.45 kg and the dry matter supply through it will be 7.6 kg (8.45 x 0.9).

Answer: We have to fed concentrate mixture 7 kg and wheat straw may be offer to the animal *ad-libitum* so the animal will consume the wheat straw as per its dry matter intake capacity and it will definitely fulfill the TDN requirements of the animal.

Exercise No. 23

Nutrient Requirements and Formulation of Ration for Sheep

Indian breeds of sheep have much lower adult weight (30-40kg) and growth rate in comparison to many foreign sheep breeds (70kg). Hence the requirements are lesser then foreign breeds.

1. Nutrient requirements for maintenance

On the basis of several balance studies and digestibility trials the requirements for Indian sheep breeds has been developed and calculations are made on the basis of 2.73 g of DCP per kg of metabolic body weight and TDN and ME requirements are 27.3 g and 98 Kcal per kg of metabolic body weight respectively. The requirements for different body weight are given in the table below:

Table: Nutrient requirements for maintenance of adult sheep (ICAR, 1998)

Body Wt. (kg)	DM (g)	DCP (g)	TDN (g)	ME (Mcal)	Ca (g)	P(g)
20	575	28	258	0.93	1.1	0.8
25	678	33	305	1.10	1.3	0.9
30	775	38	350	1.26	1.5	1.0
35	873	43	393	1.42	1.7	1.2
40	964	48	434	1.56	1.9	1.3
45	1055	53	475	1.85	2.3	1.5
50	1140	56	513	1.92	2.3	1.5
55	1225	60	551	1.98	2.4	1.6
60	1310	65	588	2.12	2.6	1.7

2. Nutrient requirements for wool production

Requirements for wool production is little bit different than for other purposes as their will be special demand for mineral sulphur. It is a constituent of wool and contributes about 4% in the composition of wool. Extensive work at Central Sheep and Wool Research Institute (CSWRI), Avikanagar has been conducted for determining some actual data on requirements for wool production, given in the table below:

Table: Nutrient requirements of sheep for wool production (ICAR, 1998)

Body Wt. (kg)	DM (g)	DCP (g)	TDN (g)	ME (Mcal)	Ca (g)	P (g)	S (g)
20	730	40	330	1.19	1.5	1.0	1.7
25	870	47	390	1.41	1.7	1.1	2.1
30	1000	54	450	1.62	2.0	1.3	2.4
35	1100	60	500	1.80	2.2	1.5	2.6
40	1230	67	555	2.00	2.5	1.6	2.9
45	1350	73	610	2.20	2.7	1.8	3.2
50	1470	80	660	2.38	2.9	1.9	3.5
55	1580	85	710	2.56	3.2	2.1	3.8
60	1680	90	755	2.72	3.4	2.2	4.0

3. Nutrient requirements for pregnancy and lactation

Sheep are mainly raised for lambs and wool production, but not for milk production. However, high lactation performance is essential for the nutrition of the lamb during early age. This becomes all the more important if twins are nursed by the ewes. Pregnancy and lactation requirements are relevant to last 6 weeks of pregnancy and first two months of the lactation, respectively. During the third month of lactation, the prescribed requirements could be reduced by 25%. The lactation requirements during the first two months are approximately twice the requirements for maintenance at the corresponding body weight, which get reduced up to 1.5 times the maintenance during the remaining period. As per the body weight requirements are given in the table on next page.

4. Nutrient requirements for growing lambs

The growth rate up to 200 g/ day has been obtained in some experiments; hence the data are tabulated up to growth rate of 200 g/ day. The weaning weight of lambs varies from 10 to 25 kg depending upon the breed and feeding regimen during pre-weaning period. Therefore, the growth requirements have been given for lambs ranging in body weight from 10 to 30 kg.

Table: Nutrient requirements for pregnant ewes (ICAR, 1998)

Body Wt. (kg)	DM (g)	DCP (g)	TDN (g)	ME (Mcal)	Ca (g)	P(g)
25	1120	80	580	2.10	3.4	2.4
30	1280	90	665	2.40	3.8	2.7
35	1440	105	750	2.71	4.3	3.1
40	1590	115	827	2.98	5.2	3.7
45	1740	125	903	3.23	5.6	4.0
50	1880	135	978	3.53	5.6	4.3
55	2020	145	1050	3.97	6.1	4.3
60	2160	155	1121	4.04	6.5	4.6

Table: Nutrient requirements for lactating ewes (ICAR, 1998)

Body Wt. (kg)	DM (g)	DCP (g)	TDN (g)	ME (Mcal)	Ca (g)	P(g)
25	1230	95	665	2.4	6.1	4.1
30	1410	108	760	2.74	7.0	4.7
35	1580	120	855	3.08	7.9	5.3
40	1750	135	945	3.41	8.7	5.8
45	1910	150	1030	3.72	9.5	6.3
50	2070	160	1120	4.04	10.3	6.9
55	2220	170	1200	4.33	11.1	7.4
60	2372	185	1280	4.62	11.8	7.9

5. Nutrient requirements for reproduction

Although, no specific information is available on the nutrient requirements of sheep for male and female reproduction, there is evidence that deficiency of one or more nutrients in the diets may affect reproduction. It is recommended that breeding ewes could be provided 50% more nutrients over the maintenance requirements at different weights.

The practice of increasing the nutrient intake of ewes and condition prior to and during breeding is called *flushing*. Its purpose is to increase the ovulation rate and consequently the lambing rate. This special feeding of providing 25% extra nutrients, over and above their maintenance requirements at about a month or 2-3 weeks prior to breeding season and continues in the breeding season. Flushing is effective only, when the ewes are undernourished during pre-flushing period and in well nourished ewes additional flushing has no specific advantage.

Table: Nutrient requirements of growing lambs (ICAR, 1998)

Body Wt. (kg)	Rate of gain (g/d)	DM (g)	DCP (g)	TDN (g)	ME (Mcal)	Ca (g)	P(g)
10	50	300	32	195	0.54	1.5	1.0
	100	340	37	220	0.72	1.7	1.1
	150	385	42	250	0.90	1.9	1.3
15	50	450	48	290	0.81	2.2	1.5
	100	510	55	330	1.08	2.5	1.7
	150	580	60	375	1.35	2.9	1.9
	200	690	75	450	1.62	3.4	2.3
20	50	600	55	360	1.08	3.0	2.0
	100	680	63	410	1.44	3.4	2.3
	150	830	77	500	1.80	4.1	2.8
	200	1000	92	600	2.16	5.0	3.3
25	50	750	65	450	1.35	3.7	2.5
	100	850	73	510	1.80	4.2	2.8
	150	1040	89	625	2.25	5.2	3.5
	200	1250	107	750	2.71	6.2	4.2
30	50	900	74	520	1.62	4.5	3.0
	100	1035	86	600	2.16	5.2	3.4
	150	1290	107	750	2.71	6.4	4.3
	200	1550	128	900	3.25	7.7	5.2

Exercise: Formulate balanced ration for a sheep weighing 60 kg for wool production, when the available feed stuffs are wheat straw and concentrate mixture (DCP 12% and TDN 60%).

Solution: We have to find out the requirements of the sheep for wool production from tables of nutrient requirements given by ICAR, 1998. On that basis we find that the capacity of dry matter intake of the sheep for the weight is 1680 g, DCP requirement is 90 g and TDN required is 755 g.

So we have to supply the required nutrients with in the range of its dry matter intake capacity and in that also we have to keep one thing in mind that their will be only concentrate mixture to supply the DCP as wheat straw don't have DCP.

So we have to find out the amount of concentrate mixture to supply complete demand of DCP i. e. 90/12 = 750 g of concentrate mixture or equivalent to 750 x 0.9 (90/100) = 675 g of dry matter.

Now how much TDN will be supplied through this amount (750 g) of concentrate mixture = 0.75 x 0.6 = 0.450 kg or 450 g.

The remaining amount of TDN will be 755 - 450 = 305 g.

Now we have to find out the amount of wheat straw required for fulfilling the requirements of remaining TDN i. e. 305/42 = 726 g (rounded up 750 g) which will provide dry matter again 675 g.

So here it is clear that within the range of dry matter intake of sheep requirement will be fulfilled.

Answer : We have to fed both concentrate mixture and wheat straw 750 g each and we may even go for offering 750 g of concentrate mixture and *ad libitum* wheat straw to fulfillt the requirement of sheep as the animal will consume the wheat straw as per its dry matter intake capacity and it will defenitely fulfill the TDN requirements of the animal.

Exercise No. 24

Nutrient Requirements and Formulation of Ration for Goats

Goats are valued as economical milk producer and main source for meat for a large section of population. The dry matter intake varies from 3% (Meat goat) to 4-6% (Dairy goat) in different breeds of India and it is higher than that of larger farm animals.

1. Nutrient requirements for maintenance

Adult body weight of goats may vary from 15 to 45 kg in different breeds. There are also does and bucks that weight up to 60kg. Nutrient requirements for maintenance of adult goats are, therefore given for body weights ranging from 15 to 60 kg. On an average for maintenance 76 g DM is required for each kg of metabolic body weight and similarly the requirements for DCP and TDN is 3 and 30 g, respectively.

2. Nutrient requirements for pregnant does

Since adult body weight as indicated already varies from 15 to 60 kg, the requirements for pregnant does are also given for body weights ranging from 15 to 60 kg.

Table: Nutrient requirements for maintenance of adult goats (ICAR, 1998)

Body Wt. (kg)	DM (g)	DCP (g)	TDN (g)	ME (Mcal)	Ca (g)	P(g)
15	500	23	240	0.86	1.1	0.7
20	615	29	295	1.07	1.3	0.9
25	730	34	350	1.26	1.6	1.1
30	830	39	400	1.45	1.8	1.2
35	940	44	450	1.62	2.1	1.4
40	1040	48	500	1.79	2.3	1.5
45	1125	53	540	1.96	2.5	1.7
50	1230	57	590	2.12	2.7	1.8
55	1315	62	630	2.28	2.9	1.9
60	1410	66	675	2.43	3.1	2.1

Table: Nutrient requirements for pregnant does (ICAR, 1998)

Body Wt. (kg)	DM (g)	DCP (g)	TDN (g)	ME (Mcal)	Ca (g)	P(g)
15	700	42	385	1.38	2.1	1.4
20	865	52	475	1.72	2.6	1.7
25	1025	62	564	2.03	3.1	2.1
30	1170	71	645	2.33	3.5	2.3
35	1320	80	725	2.62	4.0	2.7
40	1460	88	802	2.89	4.4	2.9
45	1590	96	875	3.16	4.8	3.2
50	1725	104	984	3.42	5.2	3.5
55	1850	112	1018	3.67	5.5	3.7
60	1975	120	1086	3.92	5.9	3.9

3. Nutrient requirements for lactation

The requirements have been given on a production level of 0.5 and 1.0 kg per day. The nutrient supply should be provided according to the milk production potential of the goats. The recommended lactation requirement includes the maintenance requirements at different body weights. On an average the lactation requirement is 45 g DCP and 345 g TDN per kg of 4% FCM over and above the maintenance requirements.

Table: Nutrient requirements for lactating does (ICAR, 1998)

Body Wt. (kg)	Milk yield (kg)	DM (g)	DCP (g)	TDN (g)	ME (Mcal)	Ca (g)	P(g)
20	0.5	865	51	468	1.69	4.3	2.9
	1.0	1185	74	640	2.31	5.9	3.9
25	0.5	968	56	523	1.89	4.8	3.2
	1.0	1290	79	695	2.51	6.4	4.3
30	0.5	1060	61	573	2.07	5.3	3.5
	1.0	1380	84	745	2.69	6.9	4.6
35	0.5	1155	66	623	2.25	5.8	3.9
	1.0	1470	89	795	2.87	7.3	4.9
40	0.5	1245	70	673	2.43	6.2	4.1
	1.0	1565	93	845	3.05	7.8	5.2
45	0.5	1320	75	713	2.57	6.6	4.4
	1.0	1640	98	885	3.19	8.2	5.3
50	0.5	1410	79	763	2.75	7.0	4.7
	1.0	1730	102	935	3.32	8.6	5.7
55	0.5	1490	84	803	2.90	7.4	4.9
	1.0	1805	107	975	3.52	9.0	6.0
60	0.5	1570	88	848	3.06	7.8	5.2
	1.0	1890	111	1020	3.68	9.4	6.3

4. Nutrient requirements for growth

The requirements for growth have been given for body weights ranging from 10 to 25 kg for three different rates of growth. These requirements are based on feedlot experiments conducted at different centres. Growth rates up to 150 g/day have been obtained in kids under feedlot conditions.

Exercise: Formulate balanced ration for a pregnant doe weighing 50 kg, when the available feed stuffs are wheat straw and green berseem.

Solution: We have to find out the requirements of the pregnant doe weighing 50 kg from tables of nutrient requirements given by ICAR, 1998. On that basis we find that the capacity of dry matter intake of the sheep for the weight is 1725 g, DCP requirement is 104 g and TDN required is 984 g.

Table: Nutrient requirements of growing lambs (ICAR, 1998)

Body Wt. (kg)	Rate of gain (g/d)	DM (g)	DCP (g)	TDN (g)	ME (Mcal)	Ca (g)	P(g)
10	50	380	27	265	0.96	2.0	1.4
	100	510	37	355	1.28	2.7	1.8
	150	635	47	445	1.60	3.4	2.3
15	50	510	33	330	1.19	2.7	1.8
	100	645	43	420	1.51	3.5	2.3
	150	785	53	510	1.84	4.2	2.8
20	50	640	39	385	1.39	3.3	2.2
	100	790	49	475	1.71	4.1	2.7
	150	985	59	590	2.13	5.1	3.4
25	50	760	44	440	1.59	3.8	2.5
	100	915	54	530	1.91	4.6	3.0
	150	1070	64	620	2.24	5.3	3.6

So we have to supply the required nutrients with in the range of its dry matter intake capacity and in that also we have to keep one thing in mind that their will be only green berseem to supply the DCP as wheat straw don't have DCP.

So we have to find out the amount of green berseem to supply complete demand of DCP *i.e.* 104/28 = 3.75 kg green berseen will be required and it will supply a dry matter of 20 x (3.75/100) = 0.75 kg.

Now we have to find out that how much TDN will be supplied through 3.75 kg of berseem = 3.75 x 130 = 488 g.

The remaining amount of TDN will be 984 - 488 = 496 g.

Now we have to find out the amount of wheat straw required to fulfill the requirements of remaining TDN i. e. 496/420 =1.18 kg, which will supply dry matter of 1.18 x 0.9 = 1.06 kg. So the total dry matter supplied will be 0.75 + 1.06 = 1.81 kg and it is almost similar to the capacity of the doe.

Answer : We have to fed 3.75 kg green berseem and wheat straw may be offer to animal *ad-libitum* so the animal will consume the wheat straw as per its dry matter intake capacity and it will defenitely fulfill the TDN requirements of the animal.

Exercise No. 25

Method for Improving the Nutritive Value of Poor Quality Roughages (Urea ammoniation)

Poor quality roughages includes the wheat straw, commonly fed to ruminants but unable to provide any DCP. Urea ammoniation method, not only improves the nutritive value of straw in terms of DCP and TDN, but it also improves the dry matter intake of animal by improving its palatability. In this method of urea ammoniation, one quintal of wheat straw is treated with 4 kg fertilizer grade urea (4%) by dissolving in 40 litre of water (moisture 40%). The dissolved urea is sprayed over several layers of wheat straw and mixed properly and then kept under airtight condition by covering it with polythene sheet for a period of 4 weeks. With the presence of enzyme urease the urea get broken down in to ammonia and carbon di oxide. This ammonia on one hand improves the nitrogen content of the straw and on another hand alkalinity (by ammonia and by ammonium hydroxide) reacts on the lingo-cellulosic bonds to loosen their strength. So by increasing the nitrogen content of the straw it increase the nutritive value of the straw in terms of increasing the DCP value of the straw and on another hand by reducing the bonds of lingo-cellulose it increase the digestibility of crude fiber and hence it also increase the energy (TDN) content of the straw.

For conducting the laboratory demonstration of the method we will requires 100 g of wheat straw, Trey, Fertilizer grade urea (4 g), 40 ml water and polythene begs to make urea treated straw air tight.

Exercise No. 26

Feeding of Livestock during Scarcity Conditions

In the cases of drought and flood when the feed resources made unavailable, creates the problem of animals feeding. In those cases unconventional methods and materials will have to be used to fulfil the hunger and nutrient requirements of animals to sustain them.

Traditional methods of feeding

1. Use of by-products of commercial crops such as oil seed crops, grams etc. as well as drought resistant vegetation in combination with NPN source and molasses (as source of energy) for meeting the immediate nutritional requirements.
2. Certain varieties of tree leaves and extractions of inedible oil seeds can be used to meet the essential requirements of intact protein in ruminants. This proved to be successful during draught in many parts of India.
3. Large scale feeding of bagasse and molasses in combination with urea and mineral supplements may be adopted without any toxic effects. Some of the daily ration schedule that has been successfully tried is given below:

1. Ration for an adult non producing cattle (150-300 kg body weight):

Bagasse	2.0 kg
Molasses	0.4 kg
Sugarcane tops	8.0 kg (chopped)
Urea	22 g
Common salt	30 g
Sterilized bone meal	20 g
Trace mineral mixture	0.5 g / animal twice a week

2. Ration for a growing animal (50-150 kg body weight):

Bagasse	2.0 kg
Molasses	0.8 kg
Sugarcane tops	3.0 kg (chopped)
Urea	40 g
Common salt	20 g
Sterilized bone meal	20 g
Trace mineral mixture	0.5 g / animal twice a week

Note: In case of non availability of sugar cane tops, the ration should be fortified with 5000 IU of Vitamin A and replacing the tops with forest grass hay or any other available crop residues in surrounding of drought affected areas.

Procedure for treating bagasse with urea molasses

- Bales of bagasse spread in layers of 15 -20 cm thickness.
- Urea and molasses are mixed in empty drums. Approximately, 33% of water is added to reduce the dustiness of material.
- The urea molasses mixture is sprayed over the bagasse, which is turned frequently to ensure the proper mixing and soaking.
- For spraying urea - molasses mixture garden hand shower may be used.

Note: This process can be used with dry tree leaves, coarse forest grasses and such other material available in areas where bagasse is in short supply.

Other systems of feeding

1. Liquid urea molasses feeding

Liquid molasses containing 2 to 3% uniformly mixed urea and fortifies with minerals and vitamins is referred to as liquid feed. The principal of urea - molasses liquid feed preparation lies in the homogenous mixing of urea in liquid molasses.

a. The most practical and least toxic urea – molasses liquid feed contains

Sugarcane molasses	92.0%
Urea	2.5%
Fresh water	2.5%
Mineral mixture	2.0%
Common salt	1.0%

The mixture is fortified with vitablend (vitamins A & D_3) @ 25 g/ 100 kg.

Procedure for preparing urea molasses liquid feed

- Urea is completely dissolved in water and poured gradually into the tank containing molasses with simultaneous mixing using a wooden stirrer.
- Powdered salt and mineral mixture are sprinkled over the molasses while stirring to ensure uniform distribution of all the additives in the liquid molasses.
- Special attention is required for the uniform mixing of urea solution.
- During winter season, the viscosity of molasses increases and hence, heating of the urea molasses mixture is essential for through mixing.
- The undiluted urea - molasses liquid feed containing 65% and more dry matter can be safely stored for long periods.

b. Another combination of urea molasses liquid feed

Sugarcane molasses	83%
Urea	2%
Phosphoric acid	2%
Common salt	2%
Mineral mixture	1%
By pass protein (Cotton seed cake or fish meal)	10%

Precautions to be followed while feeding liquid urea – molasses diets

1. It is desired to supplement the liquid molasses diet to certain minimum quantity of roughages to maintain bulk.
2. It is advisable to introduce liquid feed gradually in the ration in about 10 to 15 days.
3. Fresh drinking water should be made available at all time for the animal on liquid diet.
4. It is good to provide mineral mixture or mineral licks free of choice in a separate container.

2. Urea molasses mineral block (UMMB)

- The National Dairy Development Board (NDDB) has started manufacturing urea-molasses mineral block.
- The production and utilization of urea molasses blocks assumes an increasingly important role in drought prone areas.
- When licked by the ruminant animal, the urea molasses blocks provide nitrogen to micro organism in the rumen and in turn improve the digestion of the straw, which is fed during disastrous conditions.
- These blocks are available commercially in packed condition so that they can be transported to the areas of necessity easily.

Composition of urea molasses mineral block

Sugarcane molasses	45%
Urea	15%
Mineral mixture	15%
Common salt	8%
Calcite powder	4%
Bentonite/guar gum	3%
Any vegetable oil cake	10%

Procedure for preparing urea molasses mineral block

- Molasses, urea, mineral mixture, salt and calcite powder are mixed with thorough stirring and boiling.

- Binder (Bentonite/guar gum) is added with the cake to the mixture and is poured into the block moulds and allowed to cool.
- Apart from cake, bran or some roughage source can also be incorporated to prepare complete feed blocks for feeding of ruminant livestock in drought prone areas.
- At present, dewormer included medicated blocks is commercially available for feeding to small ruminants.

3. Uromin licks

This uromin lick also called pashu chat contains besides urea, molasses and minerals, certain fillers like deoiled rice bran, maida (sieved flour), mustard cake, common salt and a feed binder (Bentonite/guar gum).

Composition of uromin lick

Sugarcane molasses	30%
Urea	10%
Deoiled mustard cake	10%
Deoiled rice bran	10%
Mineral mixture	12%
Common salt	10%
Maida (Sieved flour)	15%
Bentonite/guar gum	3%

Procedure for preparing uromin lick

- First, molasses and urea are heated together in a round iron utensil for about half an hour.
- By doing so, urea and molasses are converted into uromin, where urea-N bound with sugars of molasses is used efficiently by the rumen system.
- Now, all the other ingredients (premix) are mixed with uromin, while it is hot to prevent lump formation.
- The whole mixture is then pressed in the dye of a uromin lick making machine, preferably with the help of a hydraulic jack at a pressure of 10 tons psi.

- A hard uromin lick is ready in 20-30 minutes depending upon the atmospheric temperature.
- This brick shaped uromin lick will weigh around 3 kg and can be sealed in a polythene envelop for future use.

Part III

Applied Animal Nutrition (Non-Ruminants, ANN-211)

Exercise No. 27

Nutrient Requirements and Formulation of Diet for Swine

Pigs are highly prolific among meat producing livestock and also excellent to all other type of farm animals in the efficiency of converting feed into high quality pork. The Nutrient requirements expressed for the mono-gastric animals like poultry and pigs as percent of the diet or the amount per kg of the diet, since they are generally fed in groups rather than individual feeding practiced in case of cattle and buffaloes (where absolute requirements are given). The latest requirements available are NRC nutrient requirements for swine, 1998 (10th revised edition), BIS, 1986 specifications for pig feeds and ICAR, 1998 for growing and finishing swine. Similarly ARC requirements are also available. NRC requirements are on 90% DM basis and are for breeds like Large White Yorkshire, Landrace, Berkshire *etc.*, while BIS specifications are on 100% DM basis and for indigenous pigs, improved pigs and crossbred pigs.

The swine industry in India is not very much developed like poultry, dairy and wool industry. Majority of the pigs are reared as a backyard enterprise. However, in recent years a large number of piggeries have come up where pigs are commercially reared. It is therefore essential to know the nutrient requirements and principles of feeding hogs.

Very little systemic work has been conducted in India. As per the ICAR (1998) the requirements of nutrients for pigs of different body weights are given below:

Body weights (kg)	Crude Protein (%)	TDN (%)	DE (Kcal/kg)	Average daily gain (g/day)	Average daily feed intake (kg)	Feed to gain ratio
10-20	18	68	3,300	340	1.24	3.65
20-30	15	68	2.970	430	1.70	3.95
35-60	13	68	2,640	515	2.37	4.60
60-90	12	68	2,640	500	2.42	4.80

Pigs have a large intestine constituting 35 to 45% of the volume of gastrointestinal tract. Fermentation digestion takes place in caecum and colon and produces different volatile fatty acids, which meets about 25-30% of the maintenance requirements of pigs. Level of fibre in pig's diet should be below its maximum recommended levels otherwise it will reduce the digestibility of energy.

Specification of nutrient requirements for pig's feed (on DM basis) (BIS, 1986)

Characteristic	Starter or Creep ration	Grower ration	Finishers ration or
Breeder meal			
Moisture % Max	11	11	11
Crude protein %, Min	20	18	16
Ether extract %, Min	2	2	2
Crude fiber %, Max	5	6	8
Total ash %, Max	8	8	8
Acid Insoluble ash %, Max	4	4	4
ME (Kcal / kg), Min	3360	3170	3170
Calcium %, Min	0.6	0.6	0.6
Available Phosphorus %, Min	0.6	0.4	0.5
Iron (ppm), Min	100	90	80
Copper (ppm), Min	8	6	6
Manganese (ppm), Min	30	30	20
Zinc (ppm), Min	50	50	50
Common salt %, Max	0.5	0.5	0.5
Niacin (mg/kg)	17	14	10

Pantothenic acid (mg/kg)	11	10	10
Riboflavin (mg/kg)	3	2.4	2.2
Vitamin B_{12} (µg/kg)	15	11	11
Vitamin A (IU/kg)	1700	1300	1300
Vitamin D (IU/kg)	190	180	130

Nutrient requirements of breeding swine (NRC, 1988)

Characteristic	Bred gilts, sows and adult boars	Lactating gilts and sows
DE (Kcal/kg)	3340	3340
ME (Kcal/kg)	3210	3210
Crude Protein %	12	13

Creep ration (feed) is the feed given to suckling pigs behind a barrier (creep) which allows to be taken feed by piglets only. It supports faster growth rate in piglets before weaning (56 days) and started from 7 day onwards when sow's milk alone will not be sufficient to nourish the piglets. On 4th and 14th day of piglets life injection of iron dextran is essential when reared on pacca floor to avoid piglet anaemia.

Pigs are fed starter, grower and finisher diets as per the requirement. Dividing post-weaning phase into starter, grower and finisher is arbitrary and differs with the breed and growth pattern. NRC (1998) has these phases as up to 20 kg, 20-50 kg and 50 to 110 kg body weight (Table). The CP (%) and ME (Kcal/kg diet) requirements are 18 and 3250, 15 and 3260 and 13 and 3275, respectively (NRC, 1988). BIS (1986) specified only three types of feeds for pigs and these are creep/starter diet, pig growth meal and pig finisher/breeding meal (Table).

Breeding gilts/sows, boars, lactating sows and even during gestation period swine need to be fed on restricted quantity of feeds, while the sow should be *flushed* after weaning of piglets by putting her on a well balanced comparatively high protein ration for a couple of weeks before breeding is helpful for obtaining greater litter size and weight.

Nutrient requirements of swine allowed feed *ad-libitum* (90% DM) (NRC. 1988)

Intake and performance levels	**Swine Live weight (kg)**				
	1-5	**5-10**	**10-20**	**20-50**	**50-110**
Expected weight gain (g/day)	200	250	450	700	820
Expected feed intake (g/day)	250	460	950	1900	3110
Expected efficiency (feed/gain)	1.25	1.84	2.11	2.71	3.79
Metabolizable Energy (Kcal/kg)	3220	3240	3250	3260	3275
Protein %	24	20	18	15	13
Indispensable amino acids (important only)					
Lysine%	1.4	1.15	0.95	0.75	0.60
Methionine + Cystine %	0.68	0.58	0.48	0.41	0.34
Threonine%	0.80	0.68	0.56	0.48	0.40
Tryptophan%	0.20	0.17	0.14	0.12	0.10
Linoleic acid %	**0.1**	**0.1**	**0.1**	**0.1**	**0.1**
Mineral elements (% or amount /kg diet)					
Calcium %	0.90	0.80	0.70	0.60	0.50
Total phosphorus %	0.70	0.65	0.60	0.50	0.40
Available phosphorus %	0.55	0.40	0.32	0.23	0.15
Iron, mg	100	100	80	60	40
Copper, mg	6	6	5	4	3
Manganese, mg	4	4	3	2	2
Zinc, mg	100	100	80	60	50
Selenium, mg	0.3	0.3	0.25	0.15	0.10
Vitamins (amount /kg diet)					
Vitamin A (IU)	2200	2200	1750	1300	1300
Vitamin D (IU)	220	220	220	150	150
Vitamin E (IU)	16	16	11	11	11
Vitamin K (mg)	0.5	0.5	0.5	0.5	0.5
Available Niacin (mg)	20	15	12.5	10	7
Pantothenic acid (mg)	12	10	9	8	7
Riboflavin (mg)	4	3.5	3	2.5	2
Vitamin B_{12} (μg)	20	17.5	15	10	5

Nutritive value of the feed stuffs commonly fed to swine

Feed Ingredients	**Crude protein (%)**	**Digestible Energy (Kcal/kg)**
Maize grain	8.5	3530
Jowar grain	8.9	3280
Bajra grain	11.7	2700
Barley	11.5	3120
Ragi grain	11	3100
Wheat broken	16.5	3080
Deoiled rice bran (DORB)	14	3135
Rice polish	11.9	3713
Wheat bran	15.5	2370
Tamarind seed waste	15.4	2600
Cottonseed meal	41.7	2670
Ground nut cake, solvent extracted	49	3140
Niger seed cake	33.2	2600
Safflower meal, solvent extracted	22.9	2960
Sunflower cake, solvent extracted	26.2	3100
Sesame / til cake, solvent extracted	45	3385
Soybean meal, solvent extracted	44	3490
Linseed cake	38	3440
Skimmed milk powder	33.3	3845
Fish meal	43.1	2200
Meat meal	55.6	2805
Blood meal, dried	86	2980
Lucerne meal	17.4	1880
Tapioca	3	3540
Molasses	4.4	2210
Yeast brewer's dried	43.8	3295

The formulation of diet in case of swine is similar to that of poultry and it will be discussed in detail with poultry section of this book.

Exercise No. 28

Nutrient Requirements and Formulation of Diet for Poultry

The requirements for a given nutrient is the minimum quantity of that nutrient that will maintain normal growth and reproduction, when all other nutrients are supplied in adequate quantities and at the same time, prevent the development of symptoms of nutritional deficiency.

The accurate way to express the nutrient requirements is in amount per day per animal. Poultry are fed in groups and hence, it is not practicable to express the requirements in amounts per day per animal. So, they are expressed in relation to energy content of the diet as poultry eats within limits to satisfy their energy needs. However, the estimation of energy content of the diet is not possible on routine basis and the calculated values may very much from estimated values. Because of these difficulties, the nutrient requirements of poultry are expressed on unit weight basis.

Optimum production of either eggs or meat largely depends upon the adequacy of all the nutrients in poultry ration. It is therefore, essential to know the nutrient requirements of poultry raised for different purposes and at different age groups. Fortunately, the nutrient requirements of chicken are known in much detail and accurately then of other farm animals. Those published by NRC and ARC are the minimum requirements and include no margin of safety. In our country, the nutrient requirements for chicken were given by Beauro of Indian standards (BIS, 1992) and Indian Council of Agricultural Research (ICAR, 1998).

Nutrient requirements of egg type chickens

Egg type chickens are generally fed *ad-libitum*. The primary factor modifying quantity of feed consumed by chicken is dietary energy. Many international recommendations are expressed in ration to dietary energy. However, due to paucity of adequate information in India this method was not adopted and all nutrient requirements are expressed as percent or quantities per kg of feed.

White leghorns (egg type chicken) are fed in three phases. The growth has extends up to 20 weeks of age. It consists of starter phase (0-8 weeks) and grower or developer phase (9-20 weeks). Following growth the pullets come into production which may extends up to 72 weeks of age. Depending upon the stage, nutrient requirements differs and the diets are to be suitably formulated.

Nutrient requirements also vary depending upon the system of housing in which the birds are kept. There is evidence to show that cage - housed birds may need more protein, lower energy, calcium and phosphorus than deep litter birds. However, separate nutrient levels have not been given in the present recommendations.

Nutrient requirements of broiler type chickens

Broiler chickens are meat type birds specifically bred to achieve rapid growth during the first few weeks of their life to meet market demands. Presently, acceptable targets for broiler weights are 1kg in first 4-5 weeks of age and 1.3 to 1.5 kg by 6-7 weeks. Such performance can be achieved by high concentration of nutrients in rations and by promoting high feed consumption.

Broiler rations are formulated in two phases an initial rapid growth phase from day old to five weeks, followed by a phase for 1 or 2 weeks (ICAR, 1998).

Broiler breeders are meant for egg production for hatching purpose to get commercial broiler chicks. It is important the broiler breeders do not put on too much body weight or fat, so that the egg production, fertility and hatchability are not adversely affected. Hence, the growth is controlled from initial stages of breeders life, usually by two methods, viz one by initiating low protein, low amino acid regimen from third week of their life or by introducing feed restriction. Feed restriction is continued even during egg - laying periods in broiler breeder hens, unlike in white leghorn layers because the eggs for hatching must be of good quality.

Nutrient requirements of chicken (BIS, 1992)

Characteristics	Chick (0-8 wks)	Grower (8-20 wks)	Layer (Over 20 weeks)	Broiler starter (0-6 wks)	Broiler finisher (6-9 wks)	Breeder (Over 20 weeks)
Moisture % (Max)	11	11	11	11	11	11
ME (Kcal/kg) (Min)	2600	2500	2600	2800	2900	2600
CP % (Min)	20	16	18	23	20	18
CF % (Max)	7	8	8	6	6	8
AIA % (Max)	4	4	4	3	3	4
Salt % (Min)	0.6	0.6	0.6	0.6	0.6	0.6
Ca % (Min)	1	1	3	1.2	1.2	3.0
Available P (Min)	0.5	0.5	0.5	0.5	0.5	0.5
Linoleic acid % (Min)	1	1	1	1	1	1
Lysine % (Min)	0.9	0.6	0.65	1.2	1	0.65
Methionine % (Min)	0.30	0.25	0.30	0.50	0.35	0.30
Total S aa's (Min)	0.6	0.5	0.55	0.9	0.7	0.55
Manganese (ppm)	90	50	55	90	90	90
Zinc (ppm)	60	50	75	60	60	100
Iodine (ppm)	1	1	1	1	1	1
Iron (ppm)	120	90	75	120	120	90
Copper (ppm)	12	9	9	12	12	12
Vitamin A (IU/kg)	6000	6000	8000	6000	6000	8000
Vitamin D_3 (IU/kg)	600	600	1200	600	600	1200
Thiamine (mg/kg)	5	3	3	5	5	3
Riboflavin (mg/kg)	6	5	5	6	6	8
Pantothenic acid (mg/kg)	15	15	15	15	15	15
Nicotinic acid (mg/kg)	40	15	15	40	40	15
Biotin (mg/kg)	0.2	0.15	0.15	0.2	0.2	0.2
Vitamin B_{12} (µg/kg)	15	10	10	15	15	10
Folic acid (mg/kg)	1	0.5	0.5	1	1	0.5
α Tocopherol (mg/kg)	15	10	10	15	15	15
Choline (mg/kg)	1300	900	800	1400	1000	800
Vitamin K (mg/kg)	1	1	1	1	1	1
Pyridoxine (mg/kg)	5	5	5	5	5	8

Nutrient requirements of Leghorn type chicken (ICAR, 1998)

Nutrients	Chick (0-8 wks)	Grower (8-20 wks)	Layer (Over 20 weeks)	Breeder (Over 20 weeks)
Energy (ME, Kcal/kg)	2500	2500	2500	2500
CP % (Min)	22	16	18	18
Lysine %	0.8	0.6	0.65	0.65
Methionine %	0.30	0.25	0.30	0.30
Methionine + Cys (%)	0.60	0.50	0.55	0.55
Linoleic acid %	1	1	1	1
Calcium %	1	1	3.25	3.25
Available Phosphorus %	0.4	0.4	0.4	0.4
Nacl %	0.4	0.4	0.4	0.4
Manganese (ppm)	80	50	50	80
Zinc (ppm)	50	40	60	80
Iron (ppm)	80	60	60	80
Iodine (ppm)	0.3	0.3	0.3	0.3
Copper (ppm)	8	6	6	8
Vitamin A (IU/kg)	4000	4000	10000	10000
Vitamin D_3 (IU/kg)	400	400	400	400
Vitamin E (mg/kg)	20	10	10	10
Vitamin K (mg/kg)	1	1	1	1
Thiamine (mg/kg)	4	2	2	2
Riboflavin (mg/kg)	5	3	5	8
Pyridoxine (mg/kg)	4	4	4	6
Pantothenic acid (mg/kg)	15	15	10	15
Nicotinic acid (mg/kg)	35	20	20	20
Folic acid (mg/kg)	1	0.5	0.5	1
Biotin (mg/kg)	0.25	0.15	0.15	0.25
Vitamin B_{12} (μg/kg)	10	5	6	6
Choline (g/kg)	1.3	0.9	1.3	1.3

Feed composition tables

The average nutrient content of the feed ingredients based on the analysis done previously is available in the form of published tables called feed composition tables. These tables compiled for poultry by Central Avian Research Institute (CARI), Izatnagar are revised and reproduced as shown in the tables given below:

Table: Feed composition of important poultry feeds (on DM basis)

Feed stuff	DM (%)	CP (%)	ME (Kcal/kg)	Ca (%)	P (%)
Maize grain	89.5	9.2	3309	0.25	0.40
Wheat grain	89.8	10.3	3045	0.18	0.43
Barley grain	96.4	12	2618	0.29	0.63
Oats grain	91.7	14.7	2848	0.11	0.41
Sorghum	87.3	10.3	2645	0.18	0.32
Pearl millet	89.6	12.7	2642	0.13	0.72
Rice bran (deoiled)	92.3	14.1	2235	0.37	1.80
Rice polish	91.8	12.7	2937	0.27	1.37
Wheat bran	88.9	14.7	2069	0.19	1.12
Salseed meal (deolied)	90.4	10.4	3096	0.24	0.16
Molasses	73.6	2.8	2400	1.51	0.66
Animal and vegetable fat	97.9	-	7700	-	-
Poultry manure (dehydrated)	93.5	25.7	791	3.86	1.42
Tapioca flour	90	2.9	3000	0.58	0.12
Tapioca waste	90.4	4.1	2339	0.58	0.19
Leaf meals					
Alfalfa leaf meal (Lucerne)	91.0	19.1	1777	1.83	0.45
Protein sources					
Ground nut meal (deoiled)	93.3	48.6	2328	0.31	0.67
Soybean meal	89.9	41.7	2694	0.36	0.90
Sesame meal	90.7	39.1	1882	2.46	1.42
Safflower meal (decorticated)	90.0	42.0	2039	0.40	1.25
Safflower meal (undecorticated)	95.0	22.0	1498	0.34	0.84
Sunflower cake	89.1	37.2	2230	0.43	1.14
Linseed meal	90.7	29.6	1671	0.48	0.98
Coconut meal	91.0	22.6	1190	0.23	0.66
Cotton seed meal	92.3	25.9	1556	0.52	0.86
Mustard meal	91.3	35.1	2373	0.89	1.78
Maize gluten meal	90.3	49.9	2705	0.22	0.35
Fish meal	93.8	43.1	1834	7.16	1.67
Meat meal	92.5	56.2	2319	2.68	2.06
Meat cum bone meal	94.0	53.8	2111	11.25	5.39
Blood meal	88.8	73.4	1420	0.32	0.31
Poultry by produce meal	93.0	56.4	-	3.95	1.73
Silkworm pupae meal (deoiled)	90.5	69.8	3000	0.29	0.58
Feather meal	91.0	84.6	-	0.65	0.32
Bone meal	95.5	14.6	1044	27.0	12.11
Yeast sludge (dried)	93.0	45.0	1848	0.10	0.41
Mahua cake	90.0	19.4	2169	0.20	0.80
Liver residue meal	90.9	65.4	3000	0.54	1.35
Guar meal	89.7	42	1893	0.54	0.70

Maximum levels of inclusion of feed ingredients

Each ingredient has its maximum level of inclusion in the diet. Inclusion of an ingredient beyond the maximum level may include imbalance of nutrients, difficulty in feed formulation and may reduce the performance of birds due to the presence of anti nutritional factors beyond tolerance level. Cost is also one of the factors imposing an upper limit on the inclusion of feed ingredients. The table shown below indicates the maximum level of inclusion of common feed ingredients in poultry diets.

Maximum levels of inclusion of feed ingredients in poultry diets

Ingredients	Inclusion level (%)
Energy sources	
Maize	60
Wheat	50
Barley	20-40
Oats	10-20
Sorghum (dark variety)	10-20
Sorghum (white variety)	30-40
Pearl millet	50
Ragi	10-20
Bajra	10-20
Rice	40
Rice bran	10-20
Rice polish	25-40
Rice polish (deoiled)	10-20
Wheat bran	10-15
Maize bran	10
Maize grit	10
Salseed meal (deolied)	3-5
Molasses	0-5
Animal and vegetable fat	10
Poultry manure meal	10
Tapioca flour	10-20
Gram chuni	10-15
Leaf meals	
Alfalfa leaf meal	5
Ground leaf meal	5
Grass meal	5

Protein sources	
Ground nut meal	40
Ground nut meal (deoiled)	20
Soybean meal	40
Sesame meal	20
Safflower meal	20
Sunflower meal	20
Linseed meal	4
Linseed meal (water treated)	20
Coconut meal	5
Cotton seed meal (de-gossypolised)	5
Mustard meal	10
Maize gluten feed	15
Maize gluten meal	15
Maize germ meal	15
Niger meal	10
Fish meal	10
Fish scrap	5
Fish soluble	3
Meat meal	10
Meat cum bone meal	5
Blood meal	3
Poultry by produce meal	5
Poultry hatchery by product meal	3
Silkworm pupae meal	6
Feather meal	2
Skim milk (dried)	3
Penicillin mycelium waste	15
Fermentation by-products	
Distillers dried soluble	10
Distillers dried grains	10
Distillers dried grains with soluble	10
Yeast (dried)	5
Yeast sludge (dried)	10
Mahua cake	8

The requirements of different nutrients may be discussed in detail as follows:

Water

Water is an essential nutrient but it is difficult to state the exact requirement. The requirements depend on environmental temperature, humidity, composition of the diet, rate of egg production etc. Generally, birds consume about twice as much water as the amount of feed consumed on a weight basis. Water should be available at all the times.

Energy

For poultry, metabolizable energy (ME) is considered. In theory, productive energy values for feedstuffs are superior to ME values because the productive values make no provision for estimating the heat increments. Evaluation of poultry feeds, on the basis of productive energy values involves tedious procedures, whereas the ME values are determined easily. Moreover, determination of ME is particularly well suited to poultry. Since undigested food residues and urinary excretions are voided together.

From the practical viewpoint, the usefulness of ME values in formulation poultry feed is dependent on two basic assumptions i. e. the ME value of an ingredient is independent of other dietary components and that the value remains relatively constant regardless of age, type or nature of production of the bird.

Protein and amino acids

The total protein requirements of poultry can be met easily. However, it is difficult to meet the requirements for essential amino acids. Essential amino acids must be provided in the diet in the right amount and proportion for better utilization for growth, egg production and hatchability. Some of the amino acids like methionine can partly compensate for deficiency of choline or vitamin B_{12}. Tryptophan may partially correct niacin deficiency.

The amino acids that are critical in the diets are arginine, threonine, lysine, methionine, cysteine and tryptophan. Threonine and tryptophan are only marginally deficient and a careful ingredient selection will avoid their deficiency. Arginine deficiency is usually not a problem in Indian diets because ground nut cake is one of the chief vegetable protein sources and which is rich in arginine. Therefore lysine and methionine are the limiting amino acids in the practical poultry diet. Lysine requirement as a percent of protein is less for egg production than for growth and a deficiency in diets for egg production is not common. Therefore, the amino acids requirements can be restricted in practical feed formulation to the requirement of lysine and methionine + cysteine in chick diets and methionine + cysteine in layer diets.

Minerals and vitamins contribute about 10% of the total cost of feed. Economisation on these, neglecting or reducing safety margin may restrict the performance of bird sometimes with heavy economic losses.

Antibiotics

Antibiotics are non nutrients, but feed compounding industry in India uses them to stimulate growth and improve feed conversion. Consequently they are classified as feed additives. Dietary supplementation of antibiotics also improves egg production. There are some reports that low concentration feeding constitutes a public health hazard by favouring the proliferation of antibiotic-resistant micro-organisms. However, considering the low intake of poultry meat it does not pose health hazard in India.

Availability and cost of feed ingredients

In computer feed formulation many nutrients can be taken into consideration as a time. However, by manual calculations only a few nutrients can be considered as a time. These include the following based on practical considerations.

Steps in diet formulation

In practical feed formulation by calculation, the essential steps are as follows. The ingredients are added to provide the nutrients per kg or 100kg. In the example given below, formulation of the diet is done per 100kg broiler starter diet.

Step I: The minor ingredients are fixed and or slack space may be left to include them letter. These includes nutrient and non-nutrient feed additives and natural feed ingredients added at a later stage in limited quantity because of cost or to provide a specific nutrient or non nutrient feed additive or balance a nutrient usually deficient in the diet.

Slake space may be left for addition of salt, calcium and phosphorus sources, supplemental amino acids and fats at a later stage to balance the diet.

Fixed minor ingredients: Trace minerals, vitamins, coccidiostate, arsenicals and antibacterial agents.

Slack space: Fat, phosphorus source, calcium source, salt and amino acids.

Fixed minor ingredients and Slack space : 5 kg

Step II: The levels of animal protein sources are fixed. These are added at fixed levels because of cost. These provide the limiting amino acids (lysine and methionine + Cystine) at higher levels than other feed ingredients.

Animal protein sources

Fish meal (43.1% CP) - 6 kg (431 g x 6 = 2.586)

Meat meal (56.2% CP) - 6 kg (562 g x 6 = 3.372)

Step III: The level of cereal by products, if any to be added may be fixed. Alternatively, the cereal by products and cereals can be mixed in certain ratio and can be added.

Cereal by product

Rice polish (12.7% CP) - 15 kg (127g x 15 = 1.905)

Step IV: Vegetable protein sources and energy sources can be added to provide the required amount of protein. Till now 32kg of ingredients were added (6 kg fish meal, 6 kg meat meal, 15 kg rice polish and 5 kg slack space) and these contributed about 7.863% CP (fish meal, 2.586, meat meal 3.372 and rice polish 1.905). Still 68kg of ingredients are to be added and 15.137% CP is to made up 23% CP.

Ground nut meal (48.6% CP) and sunflower meal (37.2% CP) in equal proportions as vegetable protein sources and maize as energy source are considered. The required protein level can be calculated by Algebraic equation or Pearson square method.

Algebraic method

Total ingredients - 68 kg

Protein - 15.137 kg

Let X represents the energy source and Y represent protein source. The protein content of protein source (equal proportion of ground nut meal and sunflower meal (i. e. 48.6 + 37.2)/2) is 42.9% and of maize is 9.2%.

X + Y = 68 1

0.092X + 0.429Y = 15.137........2

(or) multiplying equation 1 by 0.092 we get

0.092 X + 0.092 Y = 6.256.........3

Simplifying equation 1 and 2 we get

0.337 Y = 8.881

Or Y = 8.881 / 0.337 = 26.35 kg

And X = 68 - 26.35 = 41.65 kg

Pearson square method

Total ingredients - 68 kg

Protein - 15.137 kg

Protein as % = 15.137 x 100/68 = 22.26%

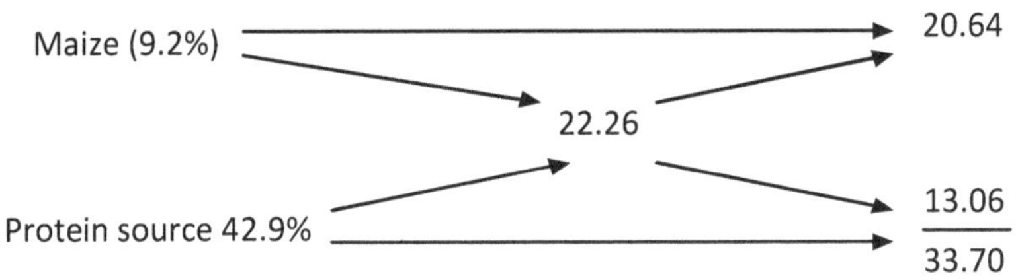

No. of parts of maize = 20.64 x 68 / 33.7 = 41.65 kg

No. of parts of protein source = 68 - 41.65 = 26.35 kg (13.175 kg each ground nut meal and sunflower meal)

Step V: The **ME** content of the diet has to be balanced. The shortfall of 75.754 kcal (2800- 2724.246 kcal) can be met by supplementation of 0.984kg fat.

Supplementation of fats is a question of economics may not be practical in all circumstances. The cost of supplemental fat may be so high as to prohibit its inclusion in poultry diets. In these cases, one has to compromise with inclusion of cereals particularly maize as energy source and the desired level of ME in these circumstances can be attained.

Step VI: The phosphorus content of the diet is calculated in terms of available phosphorus. Phosphorus from animal and inorganic sources is considered completely available, where as that from plant source is considered to be 30% available.

In this example, the deficiency of available phosphorus can be met by inclusion of 0.5 kg of tri calcium phosphate and in addition to phosphorus it will supply some amount of Ca too.

Step VII: The calcium content of the diet is calculated as 1.0225 kg (including 0.1900 kg through tri calcium phosphate). The remaining Ca content to meet the requirement can be met by supplementation of 0.5 kg limestone.

Step VIII: The sodium content usually is not calculated and 0.5% addition of commercial salt meets the requirements. However, when feed ingredients from animal sources are included, it is necessary to calculate the sodium content and accordingly salt need to be supplemented.

Step IX: Balancing the limiting amino acid in the diet is done by calculating the supply of different limiting amino acids by different feed ingredients. The animal protein sources are the richest source of the limiting amino acids, while the plant protein as well as energy sources also have their little contribution. The remaining amount of the limiting amino acids may be supplemented through their synthetic sources.

Step X: The crude fibre content of the diet will also be calculated and if the amount of fibre in diet will be higher to the maximum recommended levels then it has to be reduced, otherwise it will reduce the performance of the broilers.

Step XI: Finally a check is made for the total ingredients and also for all the nutrients if desired. If the total of the ingredients are not 100, cereal or cereal by products can be added to make it 100.

The diet can be modified by incorporating higher levels of cereals to meet energy or higher levels of vegetable or animal protein sources to meet the protein or specific amino acid requirements, if necessary. After modifications, again a check is made for all the nutrients.

Instead of a single cereal and a single vegetable protein supplement a combination of cereals or other energy sources and a combination of protein supplements may be used to reduce the cost of the diet. In addition vitamins A, D_3, E, K, B_2 and choline chloride are necessarily added. Toxin binder, antibiotic additive, growth promoter and preservatives are also added.

Example of a broiler starter diet

Ingredients	Parts	CP	ME	Ca	P
Maize	41.65	3.8318	1378.199	0.1041	0.1666
Rice polish	15.00	1.905	496.35	0.0405	0.2055
Ground nut meal	13.175	6.4031	306.714	0.0408	0.0883
Sunflower meal	13.175	4.9011	293.803	0.0567	0.1502
Fish meal	6.00	2.586	110.04	0.4296	0.1002
Meat meal	6.00	3.372	139.14	0.1608	0.1236
Animal fat	0.984	-	75.768	-	-
Tri calcium phosphate	0.5	-	-	0.1900	0.1000
Limestone	0.5	-	-	0.1900	-
Salt	0.5	-	-	-	-
Additives	2.516	-	-	-	-
Total	100	22.999	2800	1.2125	0.5069
Requirement	100	23.00	2800	1.2	0.5

Exercise No. 29

Principles of Compounding and Mixing of Feeds

Complete feed system is a system of ruminants feeding in which roughage and concentrate mixture are mixed homogenously in a predetermined ratio and processed in to pellet, briquette or block form to provide the animals a nutritionally balanced diet. Complete feed system is advantageous over conventional feed system as it reduces the cost of labour, transportation and storage of feed, besides ensuring uniform supply of nutrients and maintaining the optimum rumen environment due to the constant dietary roughage: concentrate ratio, which also avoids the selective feeding.

Feed mixing

The process of uniform distribution of different feed ingredients in the compounded feed is known as feed mixing. The feed mixing system is the heart of any feed manufacturing unit /plant. Each ingredient has its own identity until it is mixed with one or more other ingredients to make a formula feed. For an animal to grow, reproduce or lactate and remain in good health, it must receive a balanced ration that supplies nutrients at proper concentration.

There are several factors which affect uniform mixing of ingredients viz particle size and shape, density, electrostatic charge, hygroscopicity and flow ability.

Particle size is by far the most important of the ingredient factors and when ingredients have widely different particle densities, they also tend to segregate during mixing and handling.

Now a day, complete diet system is considered one of the promising methods of the animal feeding due to several reasons. Several workers developed complete feeds in the form of pellets. Use of pelleted form of complete feed is limited due to high cost of grinding and pelleting and loss of nutrients during processing. A new form of complete feed i.e. Blocks was reported in which neither grinding of straw was required nor there was loss of nutrients during processing. For the production complete feed blocks, the mixture of roughage and concentrates is compressed in a machine. To facilitate the compaction behaviour of mixture the roughage portion is mixed with a binder that is molasses. The mixing of molasses with roughage creates problems, due to large difference in bulk density and particle size. Manual mixing takes long time and cause lumps formation. Routinely used feed mixers are less efficient as the level of molasses increases in the formula hence proper care must be taken.

The most important operation in a feed mill is mixing and is the single operation that would be required in a plant to define it as a feed mill.

Theory of mixing: The aim of mixing is to disperse the ingredients of a certain assortment (called formula) so that each small unit of the whole have the same proportion of each ingredient as the original formula.

Feed mixing may require many combinations of solid and liquids. Solids are different in particle size, shape, and density, coefficient of friction, resilience and electrostatic charge. Liquids are different in density and viscosity. The objective of mixing is to scatter dissimilar parts until they are blended implying uniformity. Mixing of feeds is important, for animal performance to have uniformity having the required nutrient. Proper mixing of feed prevents toxicity when NPN substances such as urea are added. Vitamins and minerals are in traces and the distribution of their content, affect utilization of nutrients and also production.

Action of mixing: Solids have no inherent ability of their own. However, mixing of solids is often described in terms used chiefly for liquids. The following three possible mechanisms cause mixing:

1. *Convention*: The transfer of group of particles from one location in the mass to another location.
2. *Diffusion*: Distribution of particles over a freshly developed surface.
3. *Shear*: Slipping plans of solids within the mass.

Gravity and mechanical forces accomplish these actions. Improvement in mixing of solids is made by decreasing the particle size and this reduces the segregation of particles. In many cases decrease in particle size is necessary to attain a sufficient number of particles for dispersion into each portion of feed. However, when the particle size is very small as in the case of micro-ingredients, it increases the possibility of loss in normal milling equipment and certain ingredients are unstable in finely divided form and may acquire electrostatic charges. It was observed that the time to achieve good distribution in the mixing is increased with very small particles.

Feed mixing may be done manually for mixing small quantity of ground feeds and mechanically when large quantities are required to be mixed at a big livestock farm or in a feed compounding factory. Feed mixing is one of the most important steps in the manufacturing of compounded feeds.

Hand mixing

This method is used by the farmers and small scale local feed compounding units. Such feeds are mostly prepared from the locally available ingredients and often do not contain any mineral mixture and vitamin supplement. Grains, grain products and oil cakes are coarsely ground before mixing with wheat bran and or deoiled rice bran for preparing a concentrate mixture.

First of all required quantities of all ingredients are procured and crushed or coarsely ground as per requirements for species of animal for which it is going to be prepared. Now weighed amount of various feed ingredients are piled one over the other on a clean surface starting with the largest quantity or bulky feed forming the basal layer. Micro-ingredients like supplements and additives to be mixed in very small quantity form the top layer. The micro-ingredients generally require a diluents or carrier for ensuring uniform mixing. After this mixing of the feeds is done which get layered in the shape of pyramid starts from one side to another. The mixing process is repeated 4 to 5 times for uniform mixing of all the feed ingredients in the form of a mash.

The mixing of molasses is somewhat tedious and more so during the cold season when its thickness may be reduced by heating, which facilitates proper mixing.

Mechanical mixing

The machine used for the mixing of feed ingredients is called mixer and several designs of vertical and horizontal mixers have been used and the selection is done on the basis of efficiency of mixing the feed ingredients. All the mixers are designed to ensure proper mixing of various feed ingredients in shorter time and least expenditure.

Mixing equipment of different types, with tumbling, stirring, smearing and impact actions have shown to have the rate of mixing dependent on the properties of the materials mixed as well as the type of equipment used. The differences in performances of missing equipments are reduced when the materials have nearly the same particle size and density. The equipments are standardized periodically by using the standard testing procedure.

1. Studying the operation of the equipment with various materials.
2. Using various percentages of capacity loads.
3. Testing various mixing speeds.

Two types of mixers are generally used for mixing feeds:

Vertical mixer: These have single or double screws for elevating the material. However, single screw mixer is popular. These are relatively inexpensive and do a good job of dry mixing. Require less floor space and consume less power. These are little slower than horizontal mixers and require more time per batch and are not used in larger feed mills. Also due to opening at one place for discharge of the feed, cleaning is difficult and clean out will be to a lesser extent. These cannot be equipped with surge bins and ingredient hopper scales.

Horizontal mixer: These are used in small mills as well as large feed mills and have horizontal ribbon blenders. In this with right and left hand augers the material is conveyed from one end of the mixer to the other, while it is tumbled within the mixer. These are expensive and do good job of dry and liquid mixing and are faster than vertical mixers. These mixers have opening at several places along the bottom to aid in more rapid discharge and for cleaning internal ribbons and tub. So cleanout is generally 100% and is more efficient. These mixers consume more power and short leg mixers can be ceiling mounted, requiring no floor space. In many plants these mixers are being equipped with surge bins for quick discharge and ingredient hopper scales for quick loading.

Premix batch mixer: is a horizontal type batch mixer for mixing micro-ingredients and have same speed and characteristics of a standard horizontal mixer.

Liquid addition: The addition of various liquids i.e. molasses, vegetable and animals fats, fish solubles, phosphoric acid, vitamin, choline chloride etc. to manufactured feeds is a standard practice. All these are valuable from a nutritional and palatability standpoint. However, the addition of any liquid can complicate feed mixing operation i. e. agglomerate formation, resulting in suboptimum micro-ingredient distribution. Whenever, liquids are added to the mixer, they should be sprayed over the entire length of the mixer. It is generally a good practice to allow the dry ingredient to mix for a short time before liquids are added. This allows the micro-ingredients to be dispersed throughout the mix so that any agglomerates formed will be pure micro-ingredients. In order to avoid agglomerate formation preheats the molasses to 95-100°F and fats to 140-210°F before adding to the feed.

Specially designed horizontal mixer known as continuous mixer is used for the mixing of higher proportions of molasses (30-40%) in the mixer feed.

Suggested Readings

Anonymous, 1974. Laboratory Manual for Principles of Animal Nutrition. Animal Science Department, College of Veterinary Science, Tirupati.

AOAC 1995. Official Methods of Analysis. 16th Edn., Association of Official Analytical Chemists, Washington, DC 20094.

Banerjee, G. C. 1988. Feeds and principles of Animal Nutrition 1st Revised edition, Oxford & IBH Publishing Co. Pvt. Ltd., New Delhi.

Datt, C., Niranjan, M. and Singh N. P. 2006. Analytical Techniques in Animal Nutrition. Kalyani Publishers, New Delhi- 110002.

Gupta, P. C., Khatta, V. K. and Mandal, A. B. 1992. Analytical Techniques in Animal Nutrition. Directorate of Publications. CCS HAU, Hisar 125004, Haryana.

IS: 6874 (Part I). 1970. Indian Standards. Methods of Tests for Animal Feeds and Feeding stuffs. Part I. General Methods, Indian Standards Institute, New Delhi-110002.

ISI: 1992. Indian Standards Institution (IS: 13- 74). Poultry Feed Specification.

Narahari, D. 2000: Field Methods of Feed Quality Control. Technical Bulletin. American Soybean Association.

Pathak, N. N., Kamra, D. N., Agarwal, N. and Jakhmola, R. C. 1996. Analytical Techniques in Animal Nutrition Research. International Book Distributing Co., Lucknow, India.

Ranjan, S.K. and Gopal Krishna, 1980. Laboratory Manual of Nutrition Research. Vikas Publishing House Pvt. Ltd.

Ranjan, S.K. 1991. Chemical Composition of Indian Feeds and Feeding of Farm Animal. 6th new edition, ICAR, New Delhi.

Ranjan, S.K. 1998. Nutrient Requirements of Livestock and Poultry. ICAR, New Delhi.

Sastry, V.R.B., Kamra, D.N. and Pathak, N.N. 1999. Laboratory Manual of Animal Nutrition, IVRI, Izatnagar-243122 (U.P.)

Singh, R. and Pradhan, K. 1976. Laboratory Techniques for Forage Quality Evaluation. Department of Animal Nutrition, Haryana Agricultural University, Hisar-125004.

MIX
Papier | Fördert gute Waldnutzung
FSC® C083411

Zeitfracht Medien GmbH
Ferdinand-Jühlke-Straße 7
99095 Erfurt, Deutschland
produktsicherheit@kolibri360.de